AF785496

Water and Wastewater

Izharul Haq · Ankit Pratim Goswami ·
Ajay S. Kalamdhad
Editors

Water and Wastewater

Assessment, Treatment and Management Using New Technologies

CP

Springer

Editors
Izharul Haq
Department of Civil Engineering
Dr. B. Lal Institute of Biotechnology
Jaipur, Rajasthan, India

Ankit Pratim Goswami
Department of Civil Engineering
Indian Institute of Technology Guwahati
Guwahati, Assam, India

Ajay S. Kalamdhad
Department of Civil Engineering
Indian Institute of Technology Guwahati
Guwahati, Assam, India

ISBN 978-3-031-94252-5 ISBN 978-3-031-94253-2 (eBook)
https://doi.org/10.1007/978-3-031-94253-2

Jointly published with Capital Publishing Company
The print edition is not for sale in India, Sri Lanka, Pakistan, Nepal, the Maldives, Bhutan, Bangladesh and Afghanistan. Customers from India, Sri Lanka, Pakistan, Nepal, the Maldives, Bhutan, Bangladesh and Afghanistan please order the print book from: Capital Publishing Company

© Capital Publishing Company, New Delhi, India 2025

This work is subject to copyright. All rights are solely and exclusively licensed by the Publisher, whether the whole or part of the material is concerned, specifically the rights of translation, reprinting, reuse of illustrations, recitation, broadcasting, reproduction on microfilms or in any other physical way, and transmission or information storage and retrieval, electronic adaptation, computer software, or by similar or dissimilar methodology now known or hereafter developed.
The use of general descriptive names, registered names, trademarks, service marks, etc. in this publication does not imply, even in the absence of a specific statement, that such names are exempt from the relevant protective laws and regulations and therefore free for general use.
The publishers, the authors, and the editors are safe to assume that the advice and information in this book are believed to be true and accurate at the date of publication. Neither the publishers nor the authors or the editors give a warranty, express or implied, with respect to the material contained herein or for any errors or omissions that may have been made. The publishers remain neutral with regard to jurisdictional claims in published maps and institutional affiliations.

This Springer imprint is published by the registered company Springer Nature Switzerland AG
The registered company address is: Gewerbestrasse 11, 6330 Cham, Switzerland

If disposing of this product, please recycle the paper.

Preface

The book, *Water and Wastewater: Assessment, Treatment and Management Using New Technologies* presents the most recent research and innovations in the essential domain of water and wastewater management. This extensive collection consolidates comments from prominent professionals and scholars actively involved in tackling the difficulties of water resource management and wastewater treatment. Due to escalating demands on water systems from population growth, urbanisation and industrial activity, sustainable water and wastewater management has emerged as a global need. This book offers a comprehensive examination of novel techniques, technology and policies aimed at addressing these urgent concerns.

This book comprises a variety of unique studies and developments in water and wastewater management and treatment. The subjects addressed encompass the creation of an economical and decentralised system for residential wastewater treatment, integrating electrolysis with a macrophyte-assisted vermifilter. A subsequent chapter explores the utilisation of waste tire-derived magnetic pyro-char composite for the cleanup of oxytetracycline antibiotics from wastewater. The application of multivariate statistics to evaluate and mitigate pollution loadings in aquatic environments is examined comprehensively.

The book elaborates on approaches for decontaminating carbon-based nanomaterials from polluted water, utilising sophisticated processes such as microemulsion and cloud point extraction. An evaluation of drinking water treatment plant efficacy and water quality via an indexing methodology offers information for enhancing treatment operations. The isolation and identification of indigenous bacteria for microbial bioremediation investigations provide significant insights into biological treatment techniques. The book emphasises the utilisation of square wave voltammetry for detecting heavy metal ions in multi-analyte water analysis with microelectrodes, demonstrating an innovative method for assessing water quality and ensuring safety.

These varied chapters demonstrate the extensive scope and profundity of contemporary study, concentrating on both pioneering technology and techniques to improve global water and wastewater management practices.

Jaipur, India	Izharul Haq
Guwahati, India	Ankit Pratim Goswami
Guwahati, India	Ajay S. Kalamdhad

About This Book

The administration of water resources and wastewater treatment is a significant global concern, especially amid rapid urbanisation, population expansion and environmental deterioration. This volume offers a thorough examination of modern approaches designed to enhance water quality and treatment efficiency, addressing the necessity for sustainable and decentralised treatment options.

The book analyses essential subjects, including biological wastewater treatment, geofiltration systems and GIS-based multi-criteria decision analysis for groundwater evaluation, utilising empirical research and case studies. It examines the efficiency of pollutant removal, water quality metrics and the function of sophisticated filtering methods in reducing contamination. The amalgamation of environmental science, engineering and policy views guarantees a comprehensive understanding of water management difficulties and solutions.

This book is intended for researchers, engineers, policymakers and graduate students, functioning as both a theoretical and practical resource for the advancement of sustainable water and wastewater management. By showcasing innovative methodologies and assessing their practical implementations, it provides significant insights into the formulation of efficient, economical and ecologically sustainable treatment techniques.

This book is a crucial addition to the area, promoting multidisciplinary research and innovation to ensure the safety and sustainability of water supplies for the future.

Contents

About the Editors

Dr. Izharul Haq is an Assistant Professor at Dr. B. Lal Institute of Biotechnology in Jaipur, India, where he specializes in environmental biotechnology, with a focus on waste management using microbial processes and evaluating the toxicity of pollutants. He completed his Ph.D. in Microbiology at the CSIR-Indian Institute of Toxicology Research, Lucknow, India, and later conducted postdoctoral research at the Indian Institute of Technology, Guwahati. Dr. Haq has authored 65 publications, holds an h-index of 20, and has over 1900 citations, reflecting his impactful contributions to the field of environmental microbiology and pollutant remediation.

Dr. Ankit Pratim Goswami is a Postdoctoral Fellow in Civil and Resource Engineering at Dalhousie University and earned his Ph.D. from IIT Guwahati. Focusing on environmental science and engineering, his research addresses water quality and sediment contamination, with notable contributions like developing the Irrigation Water Quality Index (IWQI) using information entropy to enhance evaluation accuracy. His detailed work on rivers, has advanced understanding of water quality variations and informed water management policies. With over 130 citations, his work significantly contributes to environmental science.

Dr. Ajay S. Kalamdhad a Professor in Civil Engineering at IIT Guwahati, is renowned for his contributions to environmental engineering, particularly in sustainable solid waste management. His research spans mechanical composting, vermicomposting, anaerobic digestion, solid waste analysis, composting microbiology, biosorption, and water and wastewater treatment. Dr. Kalamdhad's work has significantly advanced sustainable waste management techniques, influencing academia, policy, and corporate practices. With over 250 publications in reputed journals and conferences, his research has made a profound impact on environmental engineering and conservation.

Abbreviations

BOD	Biochemical Oxygen Demand
BOD_5	Biochemical Oxygen Demand
Ca^{2+}	Calcium
CBNMs	Carbon-based Nanomaterials
Cd	Cadmium
CFHC	Cigarette Filter Hydrochar
Cl^-	Chloride
CNT	Carbon Nanotubes
COD	Chemical Oxygen Demand
CPE	Cloud Point Extraction
CPT	Cloud Point Temperature
Cr	Chromium
CSTR	Continuously Stirred Tank Reactors
Cu	Copper
DO	Dissolved Oxygen
EC	Electrical Conductivity
F^-	Fluoride
FAHP	Fuzzy-AHP
Fe	Iron
Fe_3O_4	Iron Oxide
GO	Graphene Oxide
GPZ	Groundwater Potential Zones
H_2SO_4	Sulfuric Acid
HNO_3	Nitric Acid
K^+	Potassium
$KMnO_4$	Potassium Permanganate
MAVF	Macrophyte Assisted Vermifilter
MCE	Mixed Cellulose Ester
ME	Microemulsion
Mg^{2+}	Magnesium
Mn	Manganese
MWCNT	Multiwalled Carbon Nanotubes
Na^+	Sodium
NaCl	Sodium Chloride
NH_4^-	Ammonium
NMs	Nanomaterials
NO_3^-	Nitrate
PAM	Polyacrylamide
Pb	Lead
PCTE	Polycarbonate Track-Etched
PII	Pharmaceutical Inlet Isolates
PO_4^{3-}	Phosphate

PTFE	Polytetrafluoroethylene
RS	Remote Sensing
SEM	Scanning Electron Microscope
SiO_2	Silicon Dioxide
SO_4^{2-}	Sulfate
TA	Total Alkalinity
TDS	Total Dissolved Solids
TSS	Total Suspended Solids
TEM	Transmission Electron Microscopy
TH	Total Hardness
TiO_2	Titanium Dioxide
TKN	Total Kjeldahl Nitrogen
TN	Total Nitrogen
TP	Total Phosphorus
TPC	Tire Pyro-char
UASB	Upflow Anaerobic Sludge Blanket
UV–VIS	Ultraviolet–Visible Spectroscopy
XRD	X-Ray Diffraction
ZnO	Zinc Oxide

Geofilter: A Low-Cost and Decentralized System for Domestic Wastewater Treatment

Ankit Kumar, Anshul Kumar, and Meena Khwairakpam

1 Introduction

Water contamination and pollution are global environmental challenges of paramount importance because of their potential detrimental effects on the natural environment, food production, economic growth and human health. The World Economic Forum has determined that the three most pressing global challenges are climate change, water supply and sanitation, and terrorism. Globally, more than three billion people (almost half the global population) do not have access to responsibly managed sanitation. Furthermore, because of rising urbanization, water demand exceeds the supply. Developed countries have wastewater collection and treatment infrastructure that is in good working order; however, the situation in developing countries is extremely disturbing.

India also trails in wastewater treatment in both urban and rural areas. India has 30–40% wastewater capacity in urban areas, while no treatment facilities are available for wastewater treatment in rural areas. Hence, a low-cost decentralized treatment facility could be a boon for wastewater treatment in rural areas.

Domestic wastewater mainly consists of nutrients, pathogens, biodegradable organics, suspended particles, oil and grease as well as a high biochemical oxygen demand (BOD). Diseases such as cholera, typhoid, hepatitis, etc., spread if untreated wastewater is released into the environment. In biological wastewater treatment, living organisms remove impurities and sludge to make it suitable for surface irrigation and other industrial uses. Biological wastewater treatment converts dissolved and suspended organic pollutants into biomass and nitrogen (N_2), methane (CH_4), carbon dioxide (CO_2), and sulphur dioxide (SO_2) (Tomar & Suthar, 2011).

Geofilter is a low-maintenance and low-operation cost water treatment system that addresses the problem of domestic wastewater treatment. Geofilter consists of different bedding materials to treat wastewater. In geofilter, packing medium filters out bacteria and increases surface area for faster bio-film growth and better treatment. The primary idea behind packed bed and fluidized systems is microbial immobilisation on solid surfaces. Many bacteria and other microorganisms adhere to support matrices. Biofilms are complex multi-species colonies of bacteria in which each species fits an ecological niche based on its metabolism and form. Solid surface preconditioning depends on pH, temperature and surface (Nasr et al., 2022). Geofilters are engineered natural systems that use filter bed and microorganisms to biodegrade waste and then filter, adsorb, and strain that waste (Arora & Kazmi, 2015).

Most studies on geofiltration investigated the treatment efficacy, nutrients and pollutant removal efficiency of wastewater and stabilisation effect of wastewater by geofilter (Kumar et al., 2014; Namaldi & Azgin, 2023; Sinha et al., 2008; Tauseef et al., 2023; Tomar & Suthar, 2011).

This chapter primarily aims to manage and treat domestic wastewater with a biological compact unit called geofilter. This chapter evaluates the removal rate of physio-chemical parameters, which are pH, electrical conductivity (EC), total suspended solids (TSS), turbidity, biochemical oxygen demand (BOD), chemical oxygen demand (COD), and total Kjeldahl nitrogen (TKN).

A. Kumar (✉) · A. Kumar · M. Khwairakpam
School of Agro and Rural Technology, Indian Institute of Technology Guwahati (IITG), Guwahati, Assam, India
e-mail: ankit.kumar_93@yahoo.in

© Capital Publishing Company, New Delhi, India 2025
I. Haq et al. (eds.), *Water and Wastewater*, https://doi.org/10.1007/978-3-031-94253-2_1

2 Methodology

2.1 Design of Geofilter

The experimental setup includes one cylindrical bottle with a total height of 37 cm and a diameter of 25 cm. The reactor was filled with the following bedding materials: bottom most layer as a supporting layer, the middle layer as a balance layer, and topmost layer as an active layer. In the reactor, gravels provided for supporting and drainage are further divided into two layers: (i) an 8-cm layer of coarse gravel and (ii) a 2-cm layer of fine gravel. Brahmaputra riverbed material is used as a middle layer with a height of 10 cm, whereas immature vermicast is used as a top layer with a height of 10 cm; a freeboard of 5 cm was provided in the reactor.

After setting up the reactor, tap water was fed into the reactor for 2 weeks to wash out filter media impurities. Wastewater was collected every day from the sewage treatment plant of IIT Guwahati (IITG), Assam, India, and fed to the geofilter with a hydraulic loading rate (HLR) of 1 $m^3/m^2/d$. As per the flow rate, the hydraulic retention time (HRT) of the geofilter was obtained to be 3.5 h. Wastewater sampling started after 15 days of reactor setup as geofilter was acclimatized. Sampling was done every fifth day to analyze effluent. The details and schematic diagram of geofiltration unit are shown in Table 1 and Fig. 1, respectively.

Table 1 Details of filter media in geofilter

S. no	Filter media layer	Height (cm)
1	Free board	5
2	Vermicast	10
3	Riverbed	10
4	Gravels (<2 cm)	2
5	Gravels (4–6 cm)	8

2.2 Sample Analysis

pH, electrical conductivity (EC), turbidity, total suspended solids (TSS), Biochemical oxygen demand (BOD), chemical oxygen demand (COD), and Total Kjeldahl nitrogen (TKN) were measured in the collected wastewater and geofilter effluent. The collected samples were stored in a freezer at 4 °C. The stored samples were analysed in triplicate within 24 h. Standard procedures for the examination of water and wastewater were followed for analysing all the parameters (APHA, 1995).

The geofilter removal efficiency was calculated according to Eq. (1):

$$\text{Percentage removal } (\%R) = \frac{C_i - C_o}{C_i} \times 100 \quad (1)$$

where

C_i = influent concentration (ppm) and

C_o = effluent concentration (ppm).

3 Results and Discussion

The initial characterization of the wastewater is described in Table 2. Initial characterization results demonstrated higher values for wastewater. This shows that wastewater cannot be disposed off directly without treatment. Geofiltration establishes significant changes in reducing pollutants. The initial

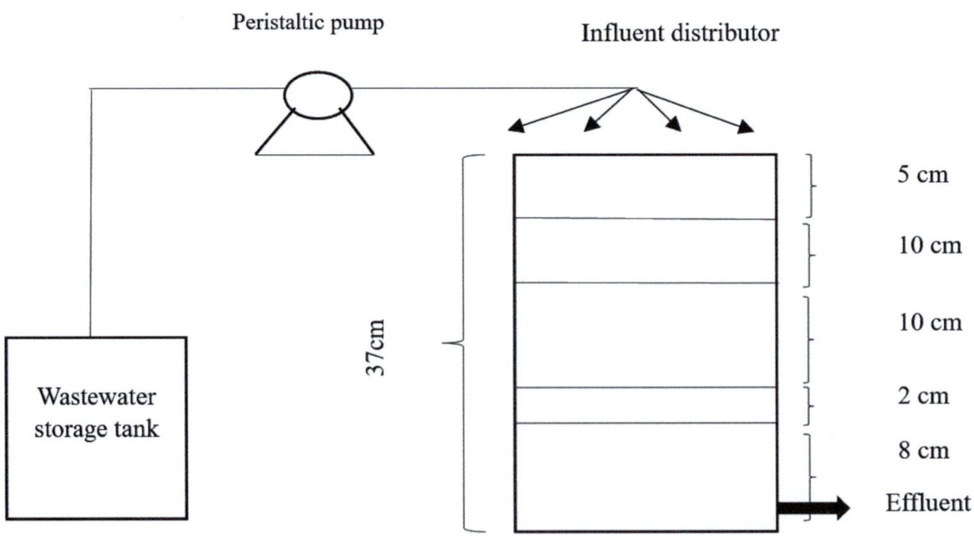

Fig. 1 Schematic diagram of a geofilter

Table 2 Initial characterisation of domestic wastewater

Parameters	Value
pH	6.45 ± 0.12
EC (mS/m)	21.69 ± 0.97
TSS (ppm)	100.59 ± 31.71
Turbidity (NTU)	211 ± 6.56
BOD (ppm)	135.33 ± 3.86
COD (ppm)	213.78 ± 14.6
TKN (ppm)	28.70 ± 2.09

Notes EC: electrical conductivity; TSS: total suspended solids; BOD: biochemical oxygen demand; COD: chemical oxygen demand; TKN: total Kjeldahl nitrogen; NTU: nephelometric turbidity unit

characterization of collected wastewater shows higher values of TSS, BOD, COD, and TOC. In untreated domestic wastewater, a typical range of BOD/COD ratio is between 0.3 and 0.8. Untreated wastewater is biologically treatable if its BOD/COD ratio is 0.5 or higher. The presence of hazardous materials in waste or the requirement of appropriate microorganisms for its stabilization is indicated by a ratio of ≤ 0.3. The BOD/COD ratio of the studied wastewater was 0.6; hence, it was easy to treat it biologically. According to Central Pollution Control Board (CPCB) standards, this wastewater couldn't be directly disposed off in rivers, for which further treatment was required.

3.1 pH Variations in Geofilter System

The trend of pH values in the study is shown in Fig. 2. The results show that pH of raw wastewater is almost neutralised in geofilter. Principal chemical components that alter the pH of wastewater are inorganic salts, organic acids, dissolved gases, and humic fractions. In order to change the pH scale of treated wastewater, certain acidic species of mineralized organic materials (carbon dioxide, nitrates, ammonia [NH_3], and organic acids) are produced if parts of organic wastewater break down in water, mostly by microbes. After the treatment process, pH of the effluent obtained after geofiltration was 6.81. Previous studies also reported an increase in pH during geofiltration (Tomar & Suthar, 2011).

3.2 Electrical Conductivity (EC) Variations in Geofilter System

Significant improvements were observed in the electrical conductivity (EC) of wastewater following treatment using a geofilter system. It's visible in Fig. 3 that geofilter has capability to reduce the electrical conductivity of domestic wastewater; EC increased initially but decreased by the 20th day onwards. The decrease in EC could be due to the absorption capacity of filter bed to adsorb ions. At the end of filtration, the removal rate of EC was 23%. These results were also supported by previous studies (Tota-Maharaj & Paul, 2015).

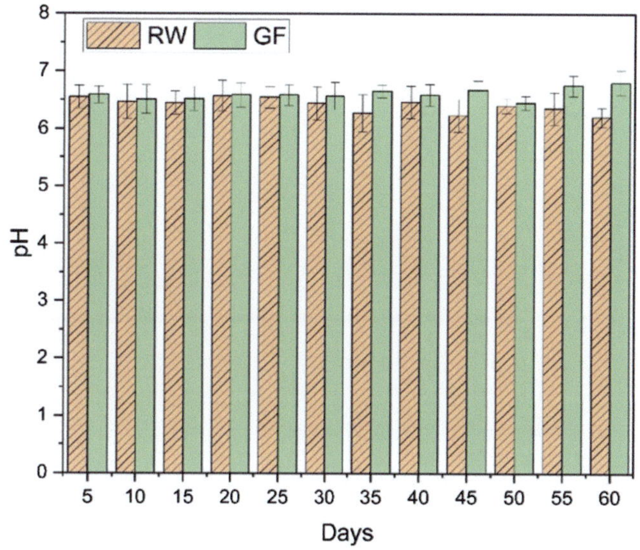

Fig. 2 Changes in pH during treatment of sewage in geofilter

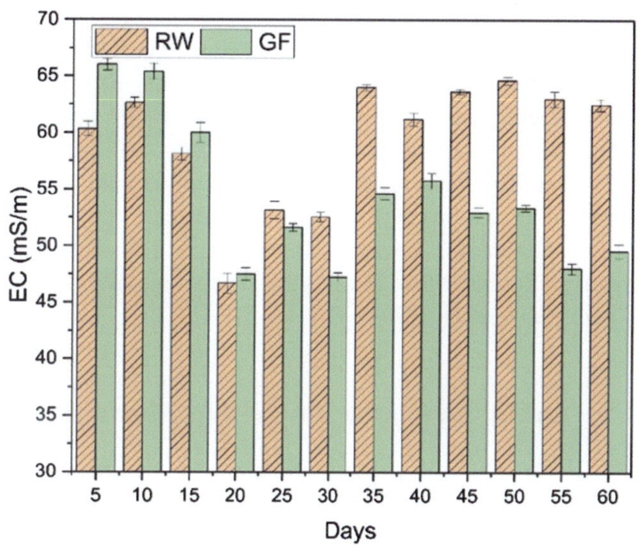

Fig. 3 Changes in electrical conductivity (EC) during treatment of sewage in geofilter

3.3 Total Suspended Solids (TSS) Variations in Geofilter System

The TSS variation in raw wastewater and treated effluent from geofilter is shown in Fig. 4. Suspended solids are removed in geofilter by trapping in filter bed voids. In this research, 60%

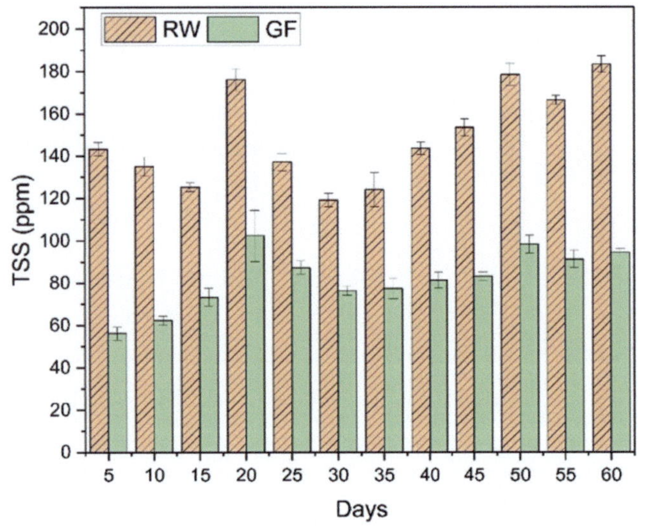

Fig. 4 Changes in total suspended solids (TSS) during treatment of sewage in geofilter

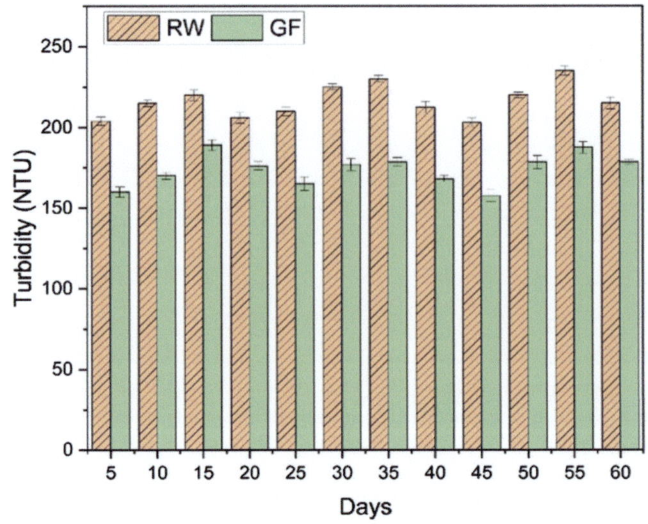

Fig. 5 Changes in turbidity during treatment of sewage in geofilter

of TSS was removed in initial days; however, as the fine particles were trapped into voids, removal efficiency of geofilter was lowered. Channelling and filter bed degradation were also visible, which could be a reason for the lower efficiency of a geofilter. The main processes involved in removal of TSS were sedimentation and filtration. The average removal of TSS using geofilter was approximately 45%. The accumulation of solids and filling of voids lead to choking of the entire system (Sinha et al., 2008).

3.4 Turbidity Variations in Geofilter System

Turbidity variations are shown in Fig. 5. The turbidity of raw wastewater ranges between 200 and 230 NTU with an average value of 217 ± 9.17 NTU whereas turbidity of geofilter effluent was in the range of 150–190 NTU, with average value of 173.10 ± 9.45 NTU. Results indicated that the turbidity removal efficiency of geofilter was 20%. Turbidity removal efficiency was low due to insufficient TSS removal efficiency and clogging of geofilter because of the trapping of solids. Adsorption of suspended particles on filter bed helps in turbidity removal (Ghasemi et al., 2020). HLR also plays an important role in removal of turbidity; therefore, turbidity removal was less because of high HLR.

3.5 Biochemical Oxygen Demand (BOD) Variations in Geofilter System

The BOD removal variations are shown in Fig. 6. It is observed that BOD was reduced in the geofilter system. The maximum BOD removal in the geofilter system was 64.5%. The BOD of

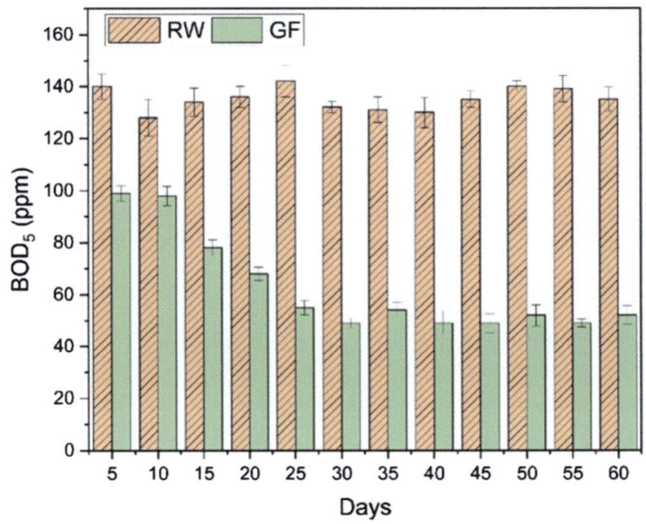

Fig. 6 Changes in biochemical oxygen demand (BOD) during treatment of sewage in geofilter

raw wastewater ranges from 130 to 143 ppm, while BOD of geofilter effluent ranges from 50 to 100 ppm. Average BOD removal efficiency of geofilter was 53.54 ± 13.76%. Aerobic microbes in geofilter degraded organic matter, resulting in the removal of BOD; however, the system got choked due to accumulation of solids and formation of colonies of bacteria and fungi on the pores of filter media (Arora et al., 2014).

3.6 Chemical Oxygen Demand (COD) Variations in Geofilter System

The geofilter system enhances COD removal from wastewater, as shown in Fig. 7. Influent COD ranges from 190 to 240 ppm, whereas COD in geofilter effluent ranges from 72

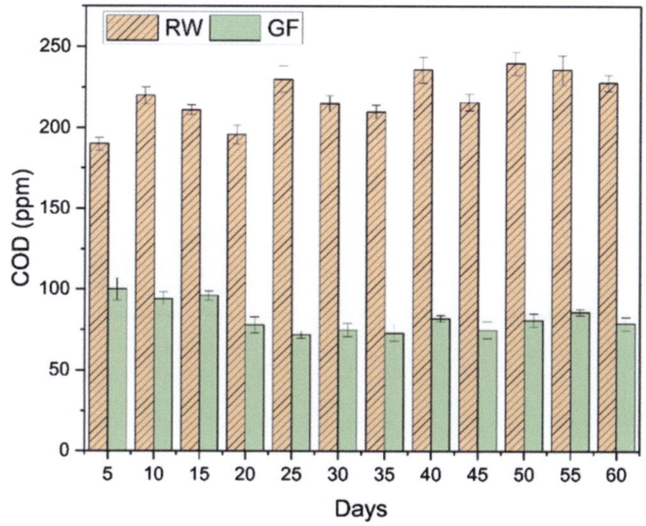

Fig. 7 Changes in chemical oxygen demand (COD) during treatment of sewage in geofilter

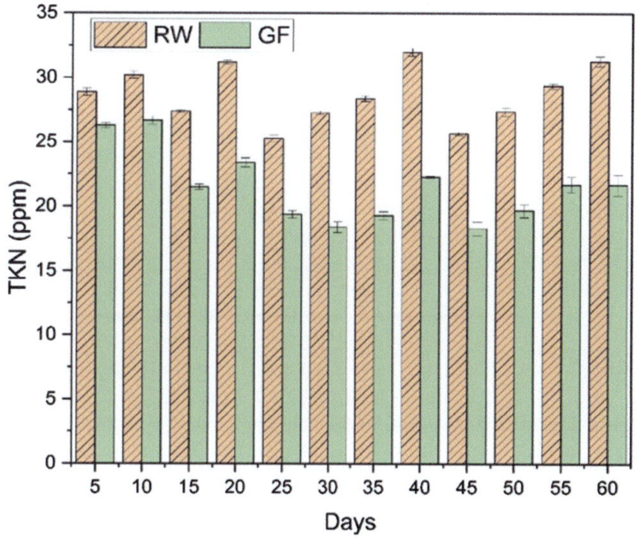

Fig. 8 Changes in total Kjeldahl nitrogen (TKN) during treatment of sewage in geofilter

to 100 ppm. The highest COD removal was 68.69%, while the average COD removal was 62 ± 5.87%, which was obtained from the geofilter treatment system. Reduced COD in effluent was due to initial filtration and adsorption through the filter, in addition to biological degradation because of microbes present in the bed (Tota-Maharaj & Paul, 2015).

3.7 Total Kjeldahl Nitrogen (TKN) Variations in Geofilter System

Variations in influent and effluent TKN during the study are shown in Fig. 8. TKN variations in raw wastewater and geofilter effluent were 25–32 ppm and 26.3–18.4 ppm, respectively. The highest TKN removal in geofilter effluent was 32.6%, with an average removal of 24.92 ± 7.32%. Ammonification, nitrification, and denitrification are the main processes involved in removal of TKN in geofilter, where microbes help to convert organic nitrogen into NH_3 and increase the nitrate content. Adsorption on ion exchange sites bound to organic matter allows microorganisms to gobble up TKN trapped on sediments in geofilter (Tomar & Suthar, 2011). The lower part of geofilter also helps in the degradation of TKN, which favours anoxic and anaerobic conditions for denitrification.

4 Conclusions

The purpose of this research was to investigate the efficacy of a geofiltration system in treating wastewater from compact residential environments, such as apartment complexes. The trial findings showed that this technique was quite effective at treating sewage. The removal efficiency of BOD (64.50%) and COD (68.69%) proved the adequate performance of microorganisms in the treatment of domestic wastewater. Removal of soluble ions was also endorsed, as EC was removed by 23%. Although these results demonstrated that the geofiltration system is effective in treating wastewater, many important questions remained unanswered, such as the effect of HLR and HRT and the microbial ecology of filter beds. To adequately fulfil the irrigation criteria for wastewater reuse in agriculture, further studies are required to optimise, operate, and improve the system's design treatment performance. Apparently, geofilter shows excellent promise as a dependable wastewater treatment technology for low-strength wastewater but is presently confronted with formidable obstacles, including a lack of funding and skilled workers, among others.

References

APHA. (1995). *American Public Health Association—Standard methods for the examination of water and wastewater* (Vol. 01).

Arora, S., & Kazmi, A. A. (2015). The effect of seasonal temperature on pathogen removal efficacy of vermifilter for wastewater treatment. *Water Research, 74*, 88–99. https://doi.org/10.1016/j.watres.2015.02.001

Arora, S., Rajpal, A., Bhargava, R., Pruthi, V., Bhatia, A., & Kazmi, A. A. (2014). Antibacterial and enzymatic activity of microbial community during wastewater treatment by pilot scale vermifiltration system. *Bioresource Technology, 166*, 132–141. https://doi.org/10.1016/j.biortech.2014.05.041

Ghasemi, S., Mirzaie, M., Hasan-Zadeh, A., Ashrafnejad, M., Hashemian, S. J., & Shahnemati, S. R. (2020). Design, operation, performance evaluation and mathematical optimization of a vermifiltration pilot plan for domestic wastewater treatment. *Journal of Environmental Chemical Engineering, 8*(1). https://doi.org/10.1016/j.jece.2019.103587

Kumar, T., Rajpal, A., Bhargava, R., & Prasad, K. S. H. (2014). Performance evaluation of vermifilter at different hydraulic loading rate using river bed material. *Ecological Engineering, 62*, 77–82. https://doi.org/10.1016/j.ecoleng.2013.10.028

Namaldi, O., & Azgin, S. T. (2023). Evaluation of the treatment performance and reuse potential in agriculture of organized industrial zone (OIZ) wastewater through an innovative vermifiltration approach. *Journal of Environmental Management, 327*. https://doi.org/10.1016/j.jenvman.2022.116865

Nasr, F. A., Abdelfadil, A. S., & ElShafai, S. A. (2022). Decentralized domestic wastewater management as unconventional water resource for agricultural purposes. *Egyptian Journal of Chemistry, 65*(5), 119–129. https://doi.org/10.21608/EJCHEM.2021.91991.4395

Sinha, R. K., Bharambe, G., & Chaudhari, U. (2008). Sewage treatment by vermifiltration with synchronous treatment of sludge by earthworms: A low-cost sustainable technology over conventional systems with potential for decentralization. *The Environmentalist, 28*(4), 409–420. https://doi.org/10.1007/s10669-008-9162-8

Tauseef, A., Mian, H. R., Khalid, H., Afzal, H., & Maqsood, H. (2023). Applicability of hybrid treatment to reduce the footprint of domestic and industrial wastewater of developing countries. *Journal of Water Process Engineering, 56*, Article 104339. https://doi.org/10.1016/j.jwpe.2023.104339

Tomar, P., & Suthar, S. (2011). Urban wastewater treatment using vermi-biofiltration system. *Desalination, 282*, 95–103. https://doi.org/10.1016/j.desal.2011.09.007

Tota-Maharaj, K., & Paul, P. (2015). Investigating the influence of geotextile layers as biofilm granular filters to treat stormwater. *Urban Water Journal, 12*(7), 559–569. https://doi.org/10.1080/1573062X.2014.916726

Multi-criteria GIS Using Fuzzy-Analytical Hierarchy Process (F-AHP) for Delineating Groundwater Potentiality Zones in the Cachar District of Assam, India

K. Zamminsion and Ajay S. Kalamdhad

1 Introduction

Groundwater resource management is crucial for urban water supply, sustainable agriculture, and healthy vegetation cover. Groundwater varies geographically as well as seasonally. With an increasing human civilisation (Gyeltshen et al., 2023), the gradual expansion of village-to-village clusters to towns and cities is observed, with changes in water consumption patterns. Rainfall and surface water have become either unavailable or scarce, especially during lean seasons, or have become unfit for human consumption due to pollution. In economically emerging nations such as India, the introduction of modern agriculture practices, and growth of industries and urban centres add up with the exponential expansion of human population, ultimately placing a significant strain on water supply regime (Gopalakrishnan et al., 2018). Furthermore, an increased reliance on groundwater in the future is anticipated as a result of increasing climate risks (Swain et al., 2022) along with inefficient management and environmental degradation (Ahmed & Sajjad, 2018) and threats to public health (Alam et al., 2022). Therefore, groundwater plays a crucial role in bridging water requirements in various cases. There is an urgent need for identification of groundwater potentiality regions, basically to give insights for allocation of appropriate recharge plans, sustainable extraction, overall revitalisation of water table as well as for the general replenishment of available groundwater for restoration of the ecosystem.

Traditionally, groundwater research methods are time-consuming and involve expensive physical procedures, which include test drillings, stratigraphy, and geophysical procedures (Bhuyan & Deka, 2022; Talukdar et al., 2022). Geospatial technologies, principally the remote sensing (RS) and geographic information system (GIS), and multi-influence factor (MIF) techniques, have proven critical in monitoring, analysing, and assessing groundwater availability and scarcity (Borah & Deka, 2023). Furthermore, these technologies aid in the assimilation of multi-parameters and production of highly accurate groundwater potential maps for large to small regions in a short period (Borah & Deka, 2023; Talukdar et al., 2022). Multiple approaches have been developed and widely employed to assign weight and rank to specific elements to perform overlay analysis on geographical information to combine multiple thematic layers to get the potential zone (Gyeltshen et al., 2023). Among them, fuzzy-based groundwater mapping models, such as the fuzzy-analytical hierarchy process (FAHP), are conventional knowledge-driven models that frequently suffer from subjectivity in case various users choose a membership function, yet they enable integration of intrinsic elements by decomposing the problem into hierarchy, assuring that the quantitative and qualitative aspects are integrated during analysis (Maity et al., 2022; Rahaman et al., 2022; Shao et al., 2020; Sresto et al., 2021). FAHP also address considerably with human subjectivity while analysing. Hence, it is considered a reliable advancement over other decision-making algorithms (Tiwari et al., 2019).

2 Study Area Description

Cachar district of Assam is located in the state's southernmost region, which is bordered on the west by neighbouring districts of Hailakandi and Karimganj, on the north by the Barali and Jayantia hill ranges, on the east by Manipur, and on the south by Mizoram. The district encompasses 3,786 km^2 with an approximate population of 2 million. Barak is the district's principal

K. Zamminsion (✉) · A. S. Kalamdhad
Centre for the Environment, Indian Institute of Technology Guwahati, Guwahati, Assam, India
e-mail: k.zamminsion@iitg.ac.in

A. S. Kalamdhad
Department of Civil Engineering, Indian Institute of Technology Guwahati, Guwahati, Assam, India

© Capital Publishing Company, New Delhi, India 2025
I. Haq et al. (eds.), *Water and Wastewater*, https://doi.org/10.1007/978-3-031-94253-2_2

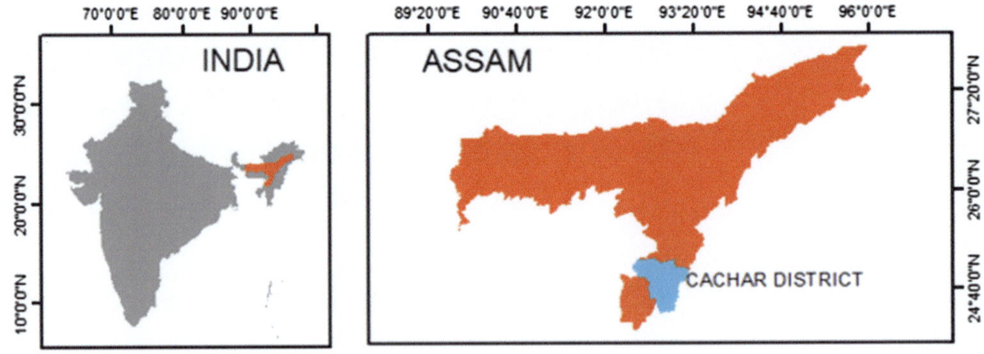

Fig. 1 Map of study area

river. Groundwater is exploited extensively for farming and anthropogenic usage.

2.1 Location and Climate

The study area was located between the longitude of 92°24′ E and 93°15′ E and latitude of 24°22′ N and 25°8′ N (Fig. 1). Apart from the Barak, which is the region's primary river, several other smaller rivers originate in Mizoram, Manipur, or the Dima Hasao district. Despite having a couple of hills scattered around it, the area primarily comprises plains. On average, over 3,000 mm of rainfall is observed at Cachar annually. The tropical wet climate features cold winters and steamy and rainy summers.

2.2 Geological Features

Geologically, the Cachar district is divided into two groups: semi-consolidated grouping with tertiary deposits dating from the Miocene to Pliocene Period from the Bhaban, Bokabil, Girujan/Tipam, Dupitila, and Dihing formations of the Miocene to Pliocene period, and the unconsolidated grouping with deposits made up of alluvial deposits dating from the Sub-Recent to Recent period. The core alluvial layers are largely composed of sand, silt, and clay, with some gravel along with coal bands. Semi-consolidated rocks are exposed as hillocks, such as shale, sandstone, ferruginous sandstone, mottle clay, pebble beds, boulder beds etc. (Ministry of Water Resources Guwahati, 2013).

2.3 Hydrogeological Features

The entire Cachar district is represented by the following three types of units: (i) unconsolidated, (ii) semi-consolidated, and (iii) consolidated (compact formation of tertiary).

(i) Surma and Dihing series of rocks, which are very compact formations;
(ii) Tipam and Dupitila formations, which are semi-consolidated; and
(iii) unconsolidated formations of alluvial deposits (Ministry of Water Resources Guwahati, 2013).

The depth to water level ranges from a few metres to 4 m below ground level in alluvial deposits, primarily in the north and south of Silchar and in the western regions, and from a few metres to 2 m below the ground level in the middle parts. The excellent repository is mainly due to the semi-consolidated Tipam sandstones in the area. The hydraulic gradient of groundwater flows from south to north-west in the southern areas and from north to south in the northern regions (Ministry of Water Resources Guwahati, 2013).

2.4 Groundwater Issues

According to the report released in 2013 by the Central Groundwater Board (Ministry of Water Resources Guwahati, 2013), the majority of groundwater is used for drinking. Owing to the lack of a sizable industry in the area, it has a low industrial water use. During heavy rainfall, floods are frequent in the area because of the disposition of clayey soil inhibiting infiltration and the lower elevation of some areas, compared to the Barak River. This leads to water logging in some regions. Nevertheless, in recent years, issues of water scarcity have become common even in some regions of the Cachar district. Hence, groundwater potentiality assessment for future management plans becomes crucial.

3 Materials and Methods

The current study considers 10 conditioning factors or parameters for the assessment of groundwater potentiality study (Table 1). Agricultural variables such as land use–land cover (LULC), normalised difference vegetation index (NDVI); hydrological variables such as lineament density (LD), drainage density, normalised difference water index (NDWI), geomorphology, slope, elevation; and meteorological

Table 1 Thematic maps, sources, and their scales

Map	Sources	Scale
LULC	Sentinel-2 10-m land use–land cover prepared by ESRI (https://www.arcgis.com/home/item.html?id=cfcb7609de5f478eb7666240902d4d3d)	10 m
Geomorphology	Geology 2 M-shape file of the Geological Survey of India (https://www.gsi.gov.in)	1:2,000,000 Topographical sheet
Lineament density	Digitised from lineament (50 K) of NRSC Bhuvan, ISRO (https://bhuvanapp1.nrsc.gov.in/thematic/thematic/index.php)	1:50,000 Topographical sheet
NDWI	Landsat-8 dataset from USGS Earth Explorer (https://earthexplorer.usgs.gov)	10 m
Drainage density	Prepared using ASTER DEM from NASA (https://earthdata.nasa.gov)	30 m
Slope	Prepared using ASTER DEM from NASA (https://earthdata.nasa.gov)	30 m
NDVI	Landsat-8 dataset from USGS Earth Explorer (https://earthexplorer.usgs.gov)	10 m
Rainfall	India Meteorological Department (https://imdpune.gov.in)	0.25 × 0.25
Geology	Geomorphology 250 K-shape file of Geological Survey of India (https://www.gsi.gov.in)	1:250,000 Topographical sheet
Elevation	Prepared using ASTER DEM from NASA (https://earthdata.nasa.gov)	30 m

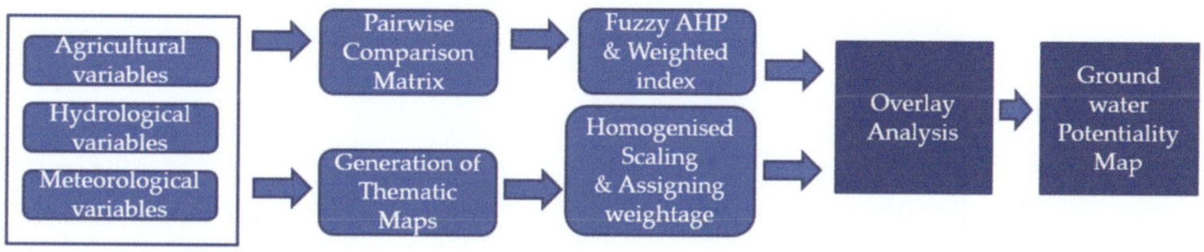

Fig. 2 Flowchart of GWP map delineation

variables including rainfall are considered for the current study as shown in Fig. 2. LULC was prepared from the Sentinel-2 dataset of 10 m spatial resolution prepared by the Environmental Systems Research Institute (ESRI). Geomorphology and geology map was clipped from the topographical sheet of geology 2 M (1:2,000,000)- and geomorphology 250 K (1:250,000)-shape file from the Geological Survey of India (GSI). Lineament density map was prepared by digitising lineament data from 50 K (1:50,000) topographical sheet of the National Remote Sensing Centre (NRSC) Bhuvan, Indian Space Research Organisation (ISRO), Government of India. NDWI and NDVI were processed from the 10 m satellite imageries of Landsat-8 datasets from the US Geological Survey (USGS) Earth Explorer. Drainage density, slope, and elevation maps were prepared using 30 m ASTER DEM from the National Aeronautics and Space Administration (NASA). Finally, rainfall data was obtained from India Meteorological Department, Pune, India.

4 Use of Remote Sensing and Geographic Information System Data for Preparation of Conditioning Parameters

Remote sensing and GIS are crucial tools for indirectly delineating groundwater potential zones and providing invaluable insights for sustainable water resource management. Geospatial techniques were applied on ArcMap 10.8 and QGIS 3.26.2 platform to generate thematic maps for 10 conditioning factors, viz.: LULC, geomorphology, lineament density, NDWI, drainage density, slope, NDVI, rainfall, geology, and elevation, where selections were made based on prior literature survey.

For the current study, the maps were procured from various agencies, whether in raw or pre-processed formats and resolutions, as depicted in Table 1, which are later homogenised during pre-processing.

5 Fuzzy-Analytical Hierarchy Process Model

The fuzzy-analytical hierarchy process is highly beneficial for multi-parametric evaluation (Saaty, 1988) if dealing with unstructured cases in geomorphology, socioeconomics, and the environment. It is a powerful decision-making tool that uses fuzzy logic and the AHP to analyse and prioritise factors impacting groundwater potential based on pairwise comparison of thematic layers using triangular fuzzy number (TFN) and Saaty's scale.

Triangular fuzzy scale (Table 2) is used for the factors that affected the potential of groundwater for being ranked with proper weights. The TFN method is chosen because of its simplicity and also because it aids in easy intuitive decision-making with clear upper- and lower-bound

Table 2 Triangular fuzzy scale (Rahaman et al., 2022)

Linguistic scale	Triangular fuzzy scale	Triangular fuzzy reciprocal scale
Absolutely more important	(5/2, 3, 7/2)	(2/7, 1/3, 2/5)
Very strongly more important	(2, 5/2, 3)	(1/3, 2/5, 1/2)
Strongly more important	(3/2, 2, 5/2)	(2/5, 1/2, 2/3)
Weakly more important	(1, 3/2, 2)	(1/2, 2/3, 1)
Equally important	(1/2, 1, 3/2)	(2/3, 1, 2)
Just equal	(1, 1, 1)	(1, 1, 1)

values. Thematic layers' weight assessments and associated grades are determined by consulting experts in the field and also by conducting thorough literature searches for an in-depth comprehension of variable ranking in various environments and across different areas.

Step A. Pairwise comparison matrix of fuzzy [F = a_{ij}] is computed as (Rahaman et al., 2022):

$$\mathrm{F} = \left\{ \begin{array}{l} (1,1,1)a_{21} - -a_{1n} \\ a_{21}(1,1,1) - -a_{2n} \\ a_{n1}a_{n2} \quad - -(1,1,1) \end{array} \right\} \tag{1}$$

where $a_{ij}\ xa_{ji} \approx 1$ and $a_{ij}\ xa_{ji} = w_i/w_j$, and all

$a_{ij} = (l_{ij}, m_{ij}, u_{ij})$, $i, j = 1, 2\ldots n$ are triangular fuzzy numbers.

Step B. Fuzzy synthetic extent (S_i) w.r.t. the ith criteria computed as:

$$S_i = \sum_{j=1}^{n} a_{ij} \left[\sum_{i=1}^{n} \cdot \sum_{j=1}^{n} a_{ij} \right]^{-1}, \tag{2}$$

where $\sum_{j=1}^{n} a_{ij}$ and $\left[\sum_{i=1}^{n} \sum_{j=1}^{n} a_{ij} \right]^{-1}$ are obtained from fuzzy addition operator.

$$\sum_{j=1}^{n} a_{ij} = \left[\sum_{i=1}^{n} l_{j,} \sum_{j=1}^{n} m_{j,} \sum_{i=1}^{n} u_j \right], \tag{3}$$

$$\left[\sum_{i=1\cdot}^{n} \sum_{j=1}^{n} a_{ij} \right] = \left[\sum_{i=1}^{n} l_{i,} \sum_{j=1}^{n} m_{j,} \sum_{i=1}^{n} u_i \right]. \tag{4}$$

Equation (4), when reversed, gives the following:

$$\left[\sum_{i=1}^{n} \cdot \sum_{j=1}^{n} a_{ij} \right]^{-1} = \left[\frac{1}{\sum_{i=1}^{n} u_j}, \frac{1}{\sum_{i=1}^{n} m_j}, \frac{1}{\sum_{i=1}^{n} l_j} \right] \tag{5}$$

Step C. Degree of possibility of two fuzzy numbers $M_2 = (l_2, m_2, u_2) \geq (l_1, m_1, u_1)$ or the intersection of two fuzzy numbers is computed as follows (Fig. 3):

$$\begin{aligned} (VM_2 \geq M_1) &= \sup_{y \geq x} \left[\min\left(\mu_{M1(x)}, \mu_{M2(y)} \right) \right] \\ &= \mathrm{hgt}\left(M_1 \cap M_2 = \mu_{M2(d)} \right) \\ &= \left[\begin{array}{c} 1, \text{ if } M_2 \geq M_1 \\ 0, \text{ if } l_2 \geq l_1 \\ \frac{(l_2 - l_2)}{(m_2 - u_2) - (m_2 - l_1)} \end{array} \right], \text{ otherwise}, \end{aligned} \tag{6}$$

where d is the ordinate between μ_{M1} & μ_{M2}.

Step D. The fuzzy number v > k may be calculated as below:

$$V(S \geq S_1, S_2, S_3, \ldots\ldots, S_k) = \min V(S \geq S_i), \quad where\, i = 1, 2, \ldots, k \tag{7}$$

Step E. Calculation of vector weight (W) with assumption that.

$d'(A_i) = \min V(S_i \geq S_k)$; i_l $k = 1, 2, 3 \ldots, n$; $i = k$; A_i ($i = 1, 2, 3, \ldots, n$) are n components.

Thus, vector weight,

$$W = \left[d'(A_1), d'(A_2), \ldots, d'(A_n) \right]^T, \tag{8}$$

Normalised non-fuzzy

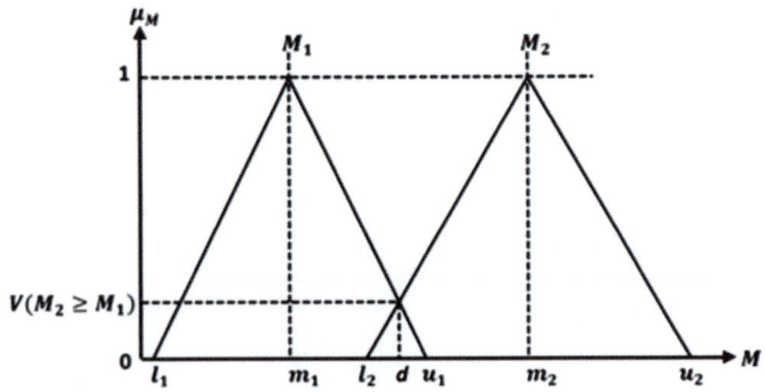

Fig. 3 Two fuzzy numbers intersection (M1) and (M2) according to Paksoy et al. (2012); and Rahaman et al. (2022)

$$Wi = [d(A_1), d(A_2), \ldots, d(A_n)]^T, \quad (9)$$

where W is a non-fuzzy number.

The resultant pairwise comparison matrices with normalised weight, *Wi*, and percentage of *Wi* for all conditioning factors are shown on Table 3.

Step F. Delineation of groundwater potentiality using FAHP.

Lastly, the groundwater potentiality (GWP) map was generated in the GIS platform by integrating all 10 thematic layers and by performing an overlay analysis in the GIS environment. The groundwater potentiality index (GWPI) was computed as follows (Maity et al., 2022; Rahaman et al., 2022; Shao et al., 2020):

$$\text{GWPI} = \sum_{i=1}^{n} * \sum_{j=1}^{n} w_i * x_j \quad (10)$$

where w_i and x_j signify the normalised weights of the ith and jth variables, respectively, while n and m represent the total number of themes and the total number of classes of the variables, respectively.

6 Consistency Ratio

A certain level of inconsistency may occur while evaluating the pairs of parameters in an AHP. Consistency ratio (CR) is a useful tool to assess strength of a model. CR works on three basic principles: relative judgement, decomposition, and priority synthesis (Shao et al., 2020). It helps in identifying and validating the results of judgment's logical consistency between pairs of parameters in an AHP (Bordoloi et al., 2023).

Consistency ratio is given as follows:

$$\text{CR} = \text{CI}/\text{RI} \quad (11)$$

where CI is the consistency index and RI is random consistency index whose values is given in Table 4 as per the number of dimensions observed (Bordoloi et al., 2023; Mahato et al., 2022).

$$\text{CI} = (\lambda_{\max} - n)/(n - 1), \quad (12)$$

where $\lambda_{\max}$ is the largest eigenvalue of the pair-wise comparison matrix, and n is the total number of parameters.

The CR must be ≤ 0.10 for AHP comparison results to be acceptable and have consistent weights (Bordoloi et al., 2023; Mahato et al., 2022). In the current study, $\lambda_{\max} = 11.399$ and CI = 0.155; therefore CR = 0.104 or 10%. Although CR is slightly higher, it is in the acceptable range. Thus, the pair of judgements is considered logically consistent.

7 Results and Discussion

Ten thematic maps were generated signifying ten conditioning factors known to influence GWP analysis in the Cachar district of Assam. These were as follows: LULC, geomorphology, lineament density, NDWI, drainage density, slope, NDVI, rainfall, geology, and elevation. FAHP technique was chosen as used by other authors for the assessment of GWP (Das & Pal, 2019; Gholami, 2022; Maity et al., 2022; Paksoy et al., 2012; Rahaman et al., 2022; Shao et al., 2020; Sresto et al., 2021). A triangular fuzzy scale was used to give weightage to pairwise comparison matrix (Maity et al., 2022). Consistency index was used for validating FAHP model, which yielded CI = 0.104, confirming the model to be logically consistent. Then, overlay analysis was exercised using ArcMap 10.8 with an evaluation scale of 1–5. The various conditioning parameters and their influence weights are given in Fig. 4.

LULC: Utilisation of land for settlement, industry, agriculture etc. is termed as land usage (Rahaman et al., 2022). Land usage and land cover play a crucial role in rainwater runoff regulation and infiltration (Sresto et al., 2021). Generally, build-up areas have lower infiltration proportions because of lesser availability of permeable surfaces, while forests and vegetation areas tend to facilitate more infiltration, as they have more capacity to retain water (Allafta et al., 2021). Changes in LULC are important drivers for groundwater changes. LULC was prepared from a Sentinel-2 dataset of 10 m resolution prepared by ESRI. It was further classified into seven classes, viz.: waterbodies, trees, grass, scrub/shrub, built-up, flooded vegetation, and bare ground. Waterbodies (2.82 area%) were assigned maximum weightage; trees (59.73%), grass (0.21%), and flooded vegetation (25.75%) were assigned moderate weightage; built-up (9.18%), bare ground (2.28%), and scrub (0.04%) were assigned minimum weightage.

Geomorphology: It offers insightful information about geology and topographic characteristics that form the foundation of the potentiality study. It greatly aids in understanding the evolution of landforms that govern the flow and occurrence of groundwater (Borah & Deka, 2023). The geomorphology shapefile of the study area was clipped in a shapefile format from the 250 K topographical sheet obtained from the Geological Survey of India. The attributes were refined and reclassified into five classes: water bodies and flood plains (5.73 area%), highly dissected hills and valleys (33.77%), alluvial plains (26.21%), pediment pediplain complex (10.97%), and low and moderately dissected hills and valleys (23.32%). Groundwater storage and recharge in the Himalayan region are facilitated by prominent structural zones, alluvial plains, and water bodies (Bordoloi et al., 2023; Gyeltshen et al., 2023); hence, they were assigned maximum weightage.

Table 3 Pair-wise comparison matrix and obtained weights for all conditioning factors

	LULC	Gm	LD	NDWI	DD	SL	NDVI	RF	Geo	Elev	WI	%
LULC	1, 1, 1	2/5, 1/2, 2/3	2/5, 1/2, 2/3	2, 5/2, 3	2/5, 1/2, 2/3	3/2, 2, 5/2	1, 3/2, 2	1, 3/2, 2	5, 5/2, 3	5, 5/2, 3	0.1209	12
Gm	3/2, 2, 5/2	1, 1, 1	2/5, 1/2, 2/3	5/2, 3, 7/2	3/2, 2, 5/2	2, 5/2, 3	5/2, 3, 7/2	2, 5/2, 3	2/5, 1/2, 5/2	5/2, 3, 7/2	0.1615	16
LD	3/2, 2, 5/2	3/2, 1, 5/2	1, 1, 1	3/2, 2, 5/2	1/2, 2/3, 1	2, 5/2, 3	1, 3/2, 2	3/2, 2, 5/2	5/2, 3, 7/2	3/2, 2, 5/2	0.1572	16
NDWI	1/3, 2/5, 1/2	2/7, 1/3, 2/5	2/5, 1/2, 2/3	1, 1, 1	1/2, 2/3, 1	1/2, 2/3, 1	1/3, 2/5, 1/2	1/2, 2/3, 1	1/3, 2/5, 1/2	3/2, 2, 5/2	0.0563	06
DD	3/2, 2, 5/2	2/5, 1/2, 2/3	1, 3/2, 2	1, 3/2, 2	1, 1, 1	5/2, 3, 7/2	2, 5/2, 3	3/2, 2, 5/2	1/2, 2/3, 1	5/2, 3, 7/2	0.1378	14
SL	2/5, 1/2, 2/3	1/3, 2/5, 1/2	1/3, 2/5, 1/2	1, 3/2, 2	2/7, 1/3, 2/5	1, 1, 1	2, 5/2, 3	2/5, 1/2, 2/3	1/2, 1, 3/2	1, 3/2, 2	0.0697	07
NDVI	1/2, 2/3, 1	2/7, 1/3, 2/5	1/2, 2/3, 1	2, 5/2, 3	1/3, 2/5, 1/2	1/3, 2/5, ½	1, 1, 1	1, 3/2, 2	1/3, 2/5, 1/2	2, 5/2, 3	0.0721	07
RF	1/2, 2/3, 1	1/3, 2/5, 1/2	2/5, 1/2, 2/3	1, 3/2, 2	2/5, 1/2, 2/3	3/2, 2, 5/2	1/2, 2/3, 1	1, 1, 1	3/2, 2, 5/2	1, 3/2, 2	0.0837	08
Geo	1/3, 2/5, 1/5	2/5, 2, 5/2	2/7, 1/3, 2/5	2, 5/2, 3	1, 3/2, 2	2/3, 1, 2	2, 5/2, 3	2/5, 1/2, 2/3	1, 1, 1	2, 5/2, 3	0.0965	10
Elev	1/3, 2/5, 1/5	2/7, 1/3, 2/5	2/5, 1/2, 2/3	2/5,1/2, 2/3	2/7, 1/3, 2/5	1/2, 2/3, 1	1/3, 2/5, 1/2	1/2, 2/3, 1	1/3, 2/5, 1/2	1, 1, 1	0.0444	04

Notes LULC: land use–land cover; Gm: geomorphology; LD: lineament density; NDWI: normalised difference water index; DD: drainage density; SL: slope; NDVI: normalised difference vegetation index; RF: rainfall; Geo: geology; Elev: elevation; WI: normalised weighted index; %: percentage of WI

Table 4 RI values with the number of dimensions of parameters chosen (Golden & Wang, 1989)

Dimension	1	2	3	4	5	6	7	8	9	10
RI	0	0	0.579	0.892	1.116	1.2368	1.332	1.395	1.454	1.488

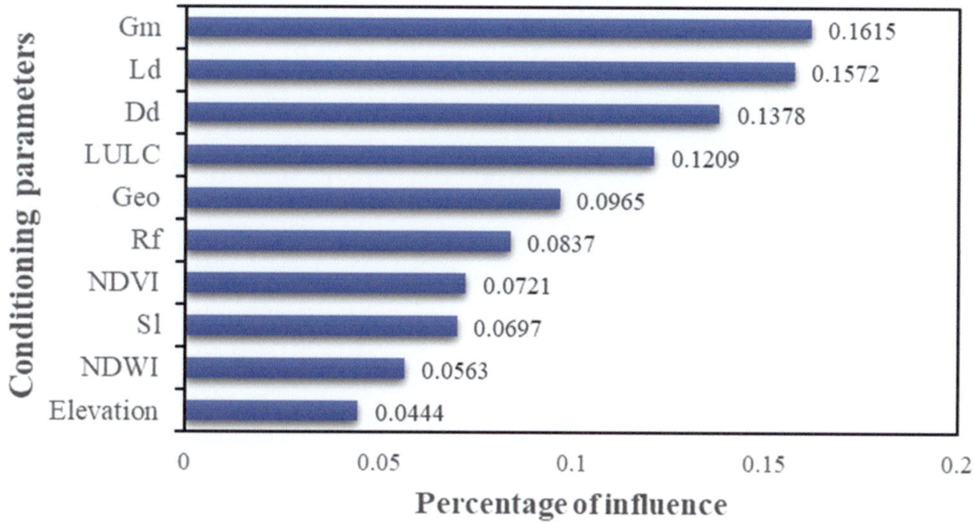

Fig. 4 Influence of each conditioning parameter by incorporating Fuzzy matrix in GWP map

Highly dissected hills and pediments were given average weightage whereas pediment pediplain complex, which were low to moderately dissected valleys and hills, was assigned minimum weightage (Fig. 5).

Lineament density: Lineaments are zones of faults and fractures that result in greater secondary porosity and permeability. As they provide pathways for the transport of groundwater, these components are essential for hydrogeology. The larger recharge potential is indicated by higher lineament density, which also increases permeability and porosity and *vice versa*. Inferentially, the groundwater potential is determined by the density of lineaments at a particular location (Gyeltshen et al., 2023; Maity et al., 2022; Rahaman et al., 2022). In the current study, the lineament density was divided into five categories, viz.: 0, 0–0.44, 0.44–0.72, 0.71–1.01, and 1.01–1.77 km/km^2, and weightage factors to them were assigned from 1 to 5, respectively (Fig. 6).

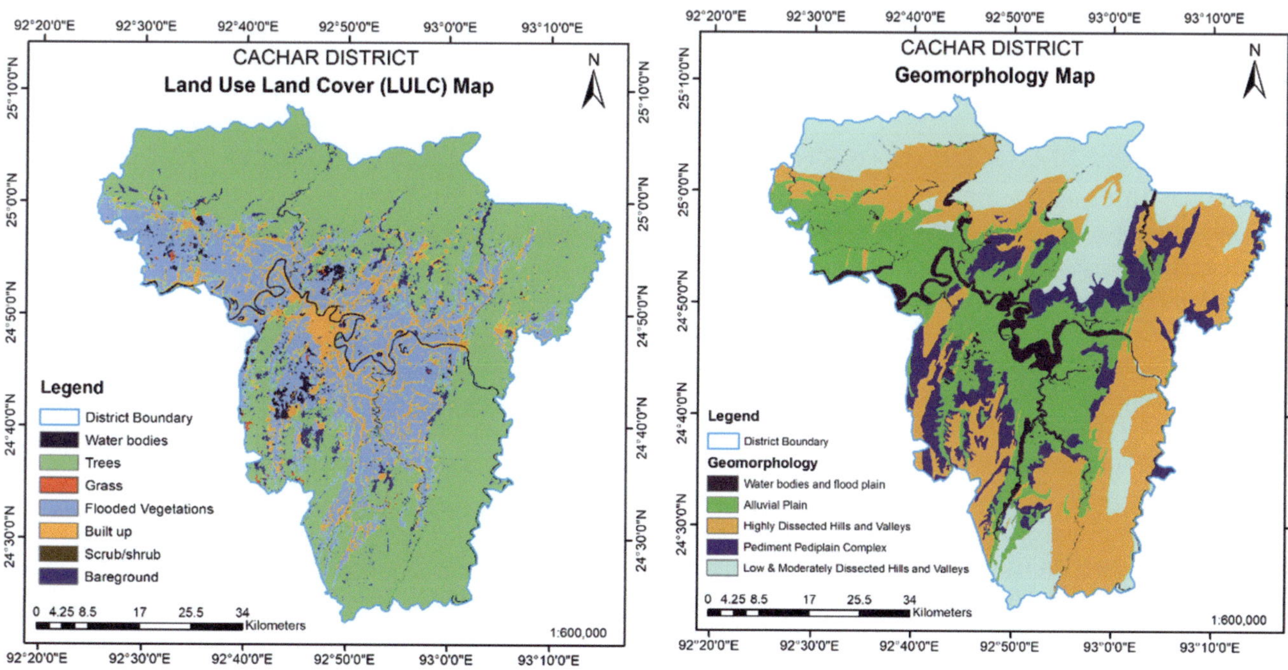

Fig. 5 LULC map (left) and Geomorphology map (right)

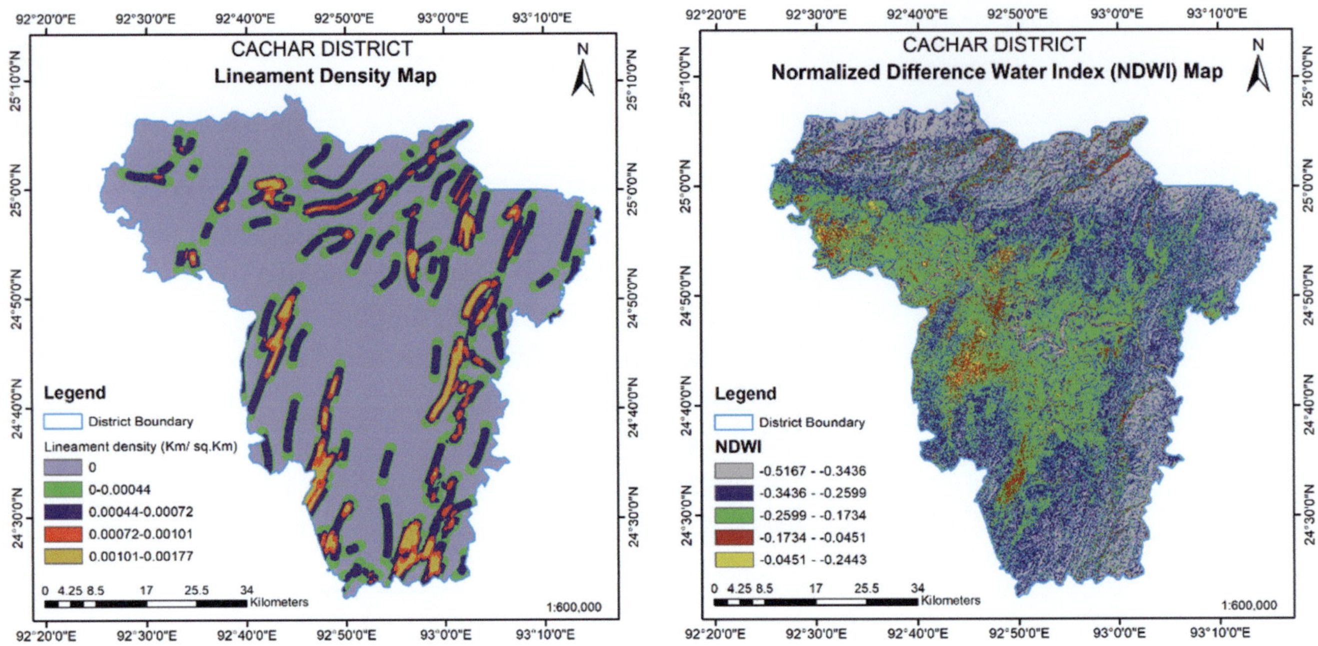

Fig. 6 Lineament density map (left) and normalized difference water index (NDWI) map (right)

NDWI: The hydrosphere is made up of surface and subsurface water, which are connected to and exchanged with each other. An intricate interaction between geohydrological, geomorphological, climatic, and landscape variables affects the naturally occurring hydrological process of surface–groundwater interaction (Bhuyan & Deka, 2022). NDWI employs green and near-infrared (NIR) bands to emphasise water bodies because water significantly absorbs light in the visible to infrared electromagnetic range. Generally, aquatic bodies have values larger than 0.5, while most bare soils ought to exhibit negative NDWI values (Gao, 1996). For the study area, the data was generated from Landsat 8 Imagery of January 2022 to obtain the least precipitation interference.

$$\text{Mathematically, NDWI} = (\text{Green} - \text{NIR band})/(\text{Green} + \text{NIR band}).$$

For the current Landsat-8 10 m satellite imagery, NDWI is equal to (Band 3 − Band 5)/(Band 3 + Band 5). The NDWI output value ranges from −0.52 to −0.25, and they are classified into five zones. Notably, higher weightage is given to higher NDWI zones because they show significant potential for groundwater recharging, compared to low NDWI zones, which show poor groundwater recharge capacity (Bhuyan & Deka, 2022). In general, aquatic bodies are associated with index values larger than 0.5. Typically, values between 0 and 0.2 correlate with built-up regions, and considerably smaller values correlate with vegetation (Fig. 6).

Drainage density: The pattern of water flow in a relief is known as a drainage system. Climate, penetration effectiveness, vegetation intensity, lithology, relief, roughness, and drainage intensity index are the factors that affect a drainage system. The ratio of total stream length to basin's or grid's area is identified as drainage density (Gyeltshen et al., 2023; Shao et al., 2020; Tiwari et al., 2019). The structural investigation of a drainage network aids in determining groundwater recharge zone, and the drainage system's quality is determined by a significant index of percolation evaluation (Maity et al., 2022). The drainage density of a study area is divided into the following five categories: 0.19–0.44, 0.44–0.63, 0.63–0.86, 0.86–1.16, and 1.16–1.53 km/km^2, with a respective area proportion of 18.86%, 34.79%, 24.99%, 11.24%, and 10.13%. Higher weightage is assigned to a higher density and vice versa (Gyeltshen et al., 2023) (Fig. 7).

Slope: Difference in slope plays a significant role in groundwater replenishment and determination of GWP zones (GWPZ). Topographically, flat regions provide more retention time for water seepage than mountainous regions with elevated slopes. Owing to their higher proportion of water renewal than steeper reliefs, sites with flatter or lower slopes are good indicators of groundwater potential zones (Shao et al., 2020). ArcGIS spatial analyst module was utilised for generating slope maps from ASTER DEM datasets of 30 m resolution. The slope of the area was divided into five categories based on the degree of inclination: 0–8, 8–22, 22–40, 40–60, and 60–180.14.

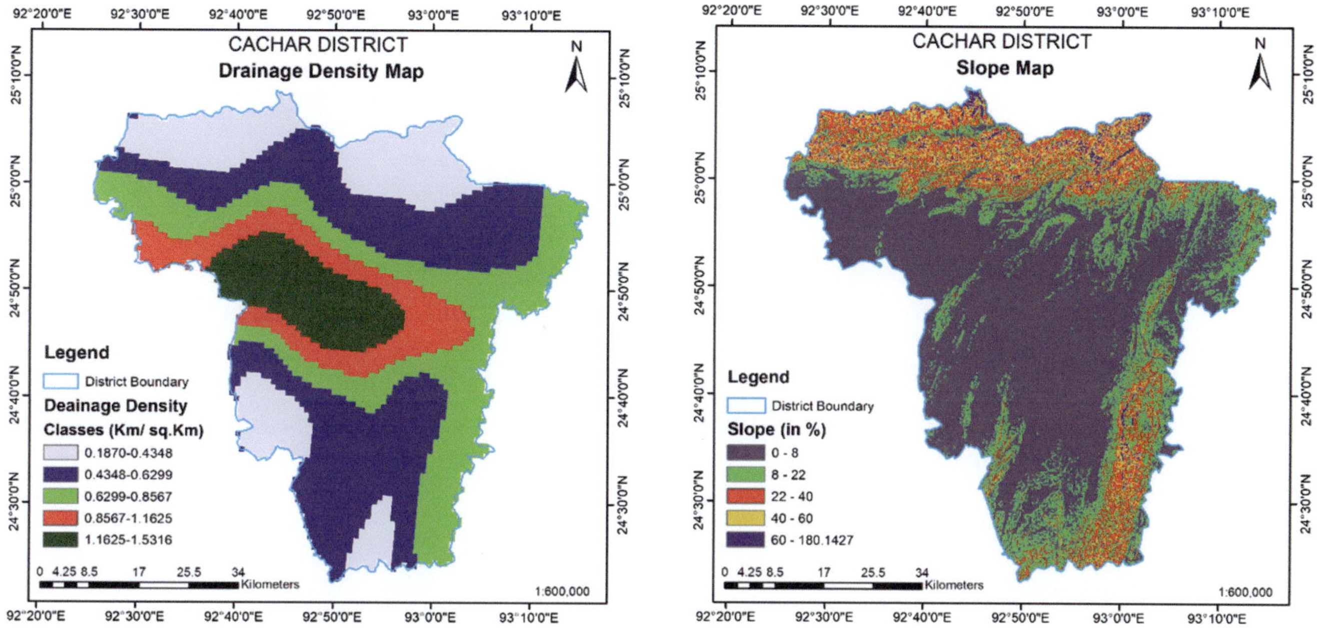

Fig. 7 Drainage density map (left) and slope map (right)

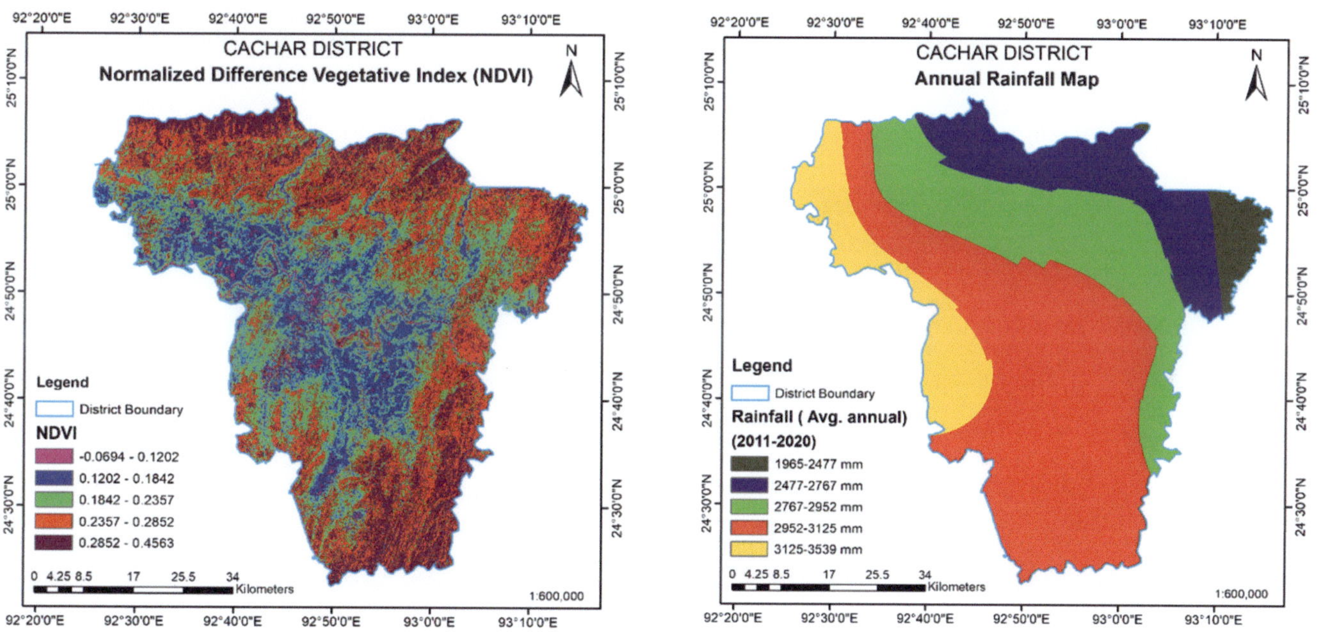

Fig. 8 NDVI map (left) and annual rainfall map (right)

NDVI: The NDVI depends on the ratio of radiation absorbed through photosynthesis. It evaluates the presence and vigour of green vegetation by comparing the reflection of visible and near-infrared light.

Mathematically, NDVI = (NIR − Red band)/(NIR + Red band). An NDVI reading of 0 indicates the absence of green plants, while NDVI of approximately + 1 (ranging from 0.8 to 0.9) represents the highest level of intensity for the observed green vegetation (Shao et al., 2020). The presence of soil moisture is also shown by NDVI based on the condition of the vegetation grown (Gyeltshen et al., 2023). For the current study, Landsat-8 10 m imagery of January 2022 was procured and processed using ArcMap. The NDVI value ranged from −0.07 to 0.46, (Fig. 8) suggesting that the vegetation was not as thick or vigorous as expected. It was divided into five categories as shown in Table 5. The NDVI categories of

Table 5 Different factors and their classes with fuzzy AHP rating, area percentage and weighted index (WI) of thematic layers

Factors	Classes	Potentiality	Fuzzy AHP rating	Area (%)	Weighted index (WI)
Land use–land cover (LULC)	Water bodies	Very good	5	2.82	0.12
	Trees	Moderate	3	59.73	
	Grass	Moderate	3	0.21	
	Flooded vegetations	Moderate	3	25.75	
	Built-up	Very poor	1	9.18	
	Scrub/shrub	Poor	2	0.04	
	Bare ground	Very poor	1	2.28	
Geomorphology	Water bodies and flood plain	Very good	5	5.73	0.16
	Alluvial plain	Good	4	26.21	
	Highly dissected hills and valleys	Moderate	3	33.77	
	Pediment pediplain complex	Very poor	1	10.97	
	Low and moderately dissected hills and valleys	Poor	2	23.32	
Lineament density (km km^{-2})	0	Very Poor	1	61.23	0.16
	0–0.00044	Poor	2	15.09	
	0.00044–0.00072	Moderate	3	17.67	
	0.00072–0.00101	Good	4	3.60	
	0.00101–0.00177	Very Good	5	2.42	
Normalised difference water index (NDWI)	−0.5167–(−0.3436)	Very poor	1	23.20	0.06
	−0.3436–(−0.2599)	Poor	2	33.37	
	−0.2599–(−0.1734)	Moderate	3	33.29	
	−0.1734–(−0.0451)	Good	4	8.19	
	−0.0451–(−0.2443)	Very good	5	1.95	
Drainage density (km km^{-2})	0.1870–0.4348	Very poor	1	18.86	0.14
	0.4348–0.6299	Poor	2	34.79	
	0.6299–0.8567	Moderate	3	24.99	
	0.8567–1.1625	Good	4	11.24	
	1.1625–1.5316	Very good	5	10.13	
Slope (°)	0–8	Very good	5	60.68	0.07
	8–22	Good	4	18.82	
	22–40	Moderate	3	12.05	
	40–60	Poor	2	6.65	
	60–180.1427	Very poor	1	1.79	
Normalised difference vegetation index (NDVI)	−0.0694–0.1202	Very good	5	3.16	0.07
	0.1202–0.1842	Poor	2	20.26	
	0.1842–0.2357	Moderate	3	27.30	
	0.2357–0.2852	Very good	5	29.63	
	0.2852–0.4563	Good	4	19.65	

(continued)

Table 5 (continued)

Factors	Classes	Potentiality	Fuzzy AHP rating	Area (%)	Weighted index (WI)
Rainfall (mm/year of 10 years avg.)	1965–2477 mm	Very poor	1	3.56	0.08
	2477–2767 mm	Poor	2	15.03	
	2767–2952 mm	Moderate	3	23.29	
	2952–3125 mm	Good	4	46.81	
	3125–3539 mm	Very good	5	11.30	
Geology	Tipam sandstone formation	Good	4	10.38	0.10
	Dupitila formation	Good	4	5.41	
	Undiff. fluvial, aeolian, coasta, and glacial sediments	Moderate	3	27.06	
	Barail group	Very poor	1	11.11	
	Surma group	Very good	5	46.04	
Elevation	−8–26	Very good	5	22.39	0.04
	26–40	Good	4	21.43	
	40–75	Moderate	3	19.25	
	75–221	Poor	2	18.77	
	221–1728	Very poor	1	18.16	

−0.07–0.12 and 0.24–0.29 were assigned maximum weightage because they covered respective water and vegetation bands, which were conducive for GWPs, while the NDVI category of 0.12–0.18 was assigned minimum weightage, as it generally covered barren land only.

Rainfall: It is a vital variable of weather in the delineation of groundwater potentiality zones. As a result, understanding the spatiotemporal phenomena of rains is necessary for the study of water resources. In the event of heavy rain, there is an elevated chance of additional groundwater. In contrast, if rainfall is minimal, the groundwater potential is reduced (Maity et al., 2022; Shao et al., 2020). Rainfall data of 10 years averaged (2011–2020) was obtained from India Meteorological Department, Pune, India. The cell resolution of raster data was interpolated from 25° × 25° to 30 m using the Kriging method of the ArcMap spatial analyst tool. The output rainfall ranges from 1,965 mm/year to 3,539 mm/year. It was further classified into five categories (Table 5), and higher rainfall values were assigned higher weightage. Although high rainfall in an area tends to increase recharge, the geological structure and high slope of area may negate the potential ground seepage capability.

Geology: It is one of the key elements to determine a region's groundwater potential, because it regulates the type and pace of groundwater percolation. Indeed, geologic formations with high porosity and permeability increase groundwater storage and yield (Bhuyan & Deka, 2022). The district's soil types range from alluvial to lateritic. Soil texture ranges from clayey loam to clay. The character of river line tracts is loamy to sandy loamy, and lateritic soil covers steep areas (Ministry of Water Resources Guwahati, 2013). Geology map is derived from 2 M shapefile of Geological Survey of India and reclassified into five categories, viz.: Tipam sandstone formation (10.38 area%); Dupitila formation (5.41%); undifferentiated fluvial, aeolian, coastal, and glacial sediments (27.06%); Barial group (11.11%), and Surma group (46.04%) and their resulted weightage was 4, 4, 3, 1, and 5, respectively (Table 5).

Elevation: Groundwater prospects are likely to be influenced by topographic elevation and governed by a variety of hydrogeological and geomorphological processes (Maity et al., 2022). The digital elevation model (DEM) procured from ASTER DEM was utilised to generate topographical elevation in this investigation. In the study area, the elevation ranged from −7 m to 1,728 m above mean sea level (msl) (Fig. 9). It was then reclassified into five categories as shown in Table 5. Lower elevation values were assigned higher weightage and vice versa, considering gravity as a major contributor.

7.1 Groundwater Potential Zone Map

In Cachar district, for groundwater potentiality study, thematic map layers were computed using weighted overlay analysis in GIS platform for yielding final output raster. The final GWP map (Fig. 10) revealed that the majority of the area was under a moderately potential region (73.05 area%), which covered largely the low lying waterlogged, plain to gentle slope, having moderate NDVI values and

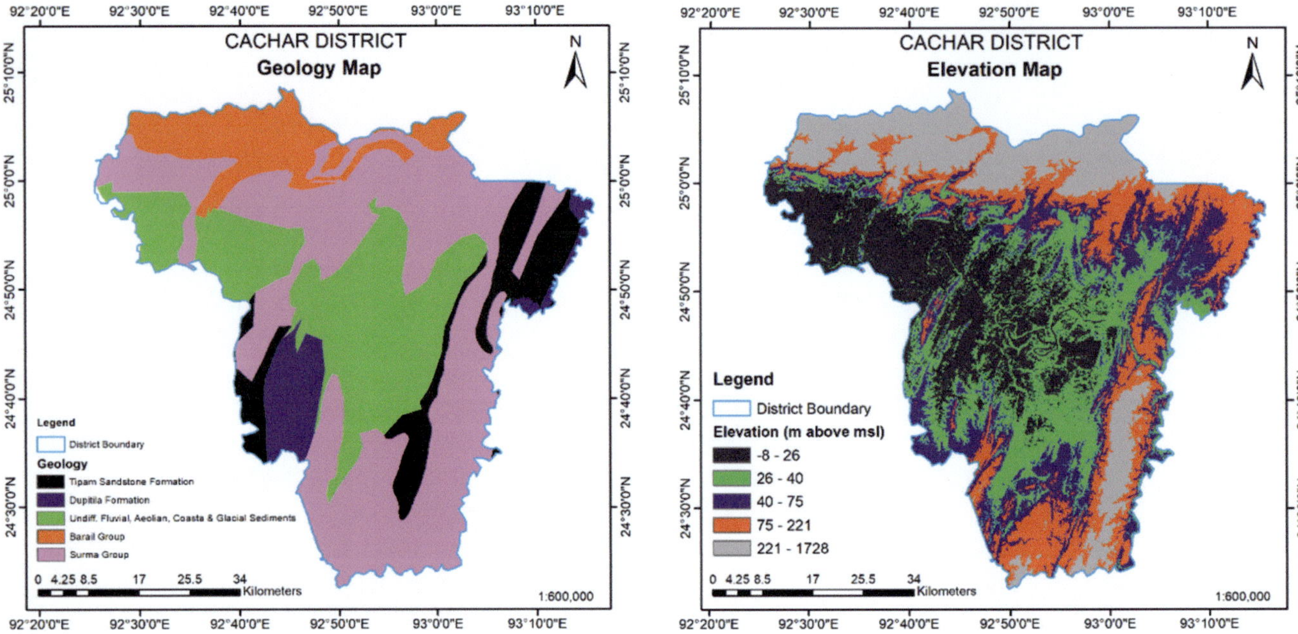

Fig. 9 Geology map (left) and elevation map (right)

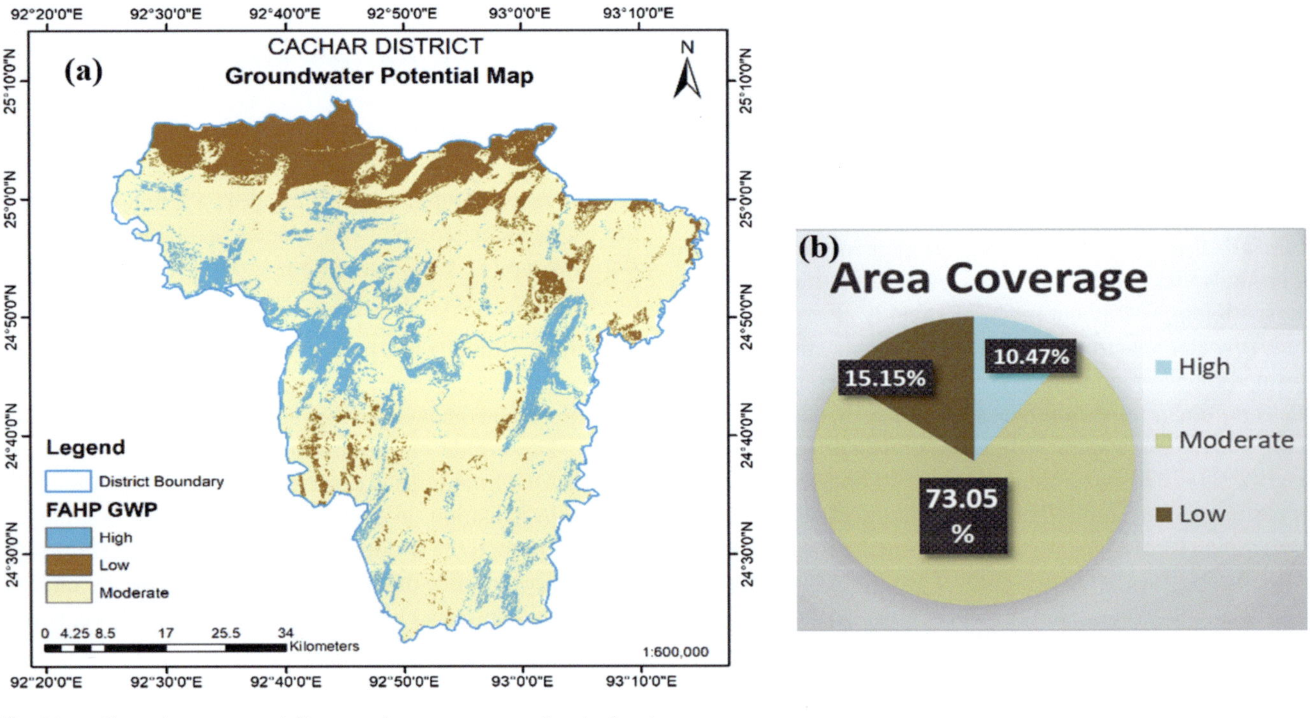

Fig. 10 **a** Groundwater potentiality map; **b** area coverage of each class in percentage

an elevation of up to 40 m above msl, while high potentiality regions (10.47 area%) were observed mostly beside the Barak River, low lying areas, and in some patches of highly dissected lineaments. Low potentiality (15.15 area%) was observed in the northern, highly elevated, high slope, and lesser drainage density areas consisting the Barail group geological formations. Surprisingly, although latter regions encompassed healthy vegetation, other above-mentioned parameters had a high impact on the overall GWP ratings. The approach proved to be an effective, convenient as well as reliable tool for GWP mapping, especially in a remotely located terrain of Cachar district. Also, as mentioned by Shao et al. (2020), the porous, unconsolidated, and flat-to-gentle hill terrains in the current study could be considered appropriate for groundwater conservations.

8 Conclusions

In the current study, FAHP model was utilised for delineating the GWP of Cachar district of Assam by incorporating 10 conditioning parameters, viz.: LULC, geomorphology, lineament density, NDWI, drainage density, slope, NDVI, rainfall, geology, and elevation. FAHP ratings were assigned and the results were validated and found logically consistent using CR, whose value was 0.104, which was in acceptable range. Thematic maps generated were classified and weighted for overlay analysis. Final GWP map generated was categorised as high, low, or moderate, with a produced area coverage of 10.47%, 15.15%, and 73.05%, respectively (Fig. 10b). High potentiality zones existed within the low-lying central regions along the Barak River, while the northern, high-altitude regions had a comparatively low groundwater potentiality. Overall, most central to southern plains were found to possess a moderate potential for groundwater recharge.

This preliminary assessment could provide local authorities, water managers, and communities a good insight into sustainable groundwater exploitation and proper recharge management plans in the zones having lower potentiality. This could be executed by construction of check-dams, along ravens and creeks, to facilitate percolation and seepage of water into the aquifer or rainwater harvesting plans as well as efficient groundwater extraction in surplus zones.

References

Ahmed, R., & Sajjad, H. (2018). Analyzing factors of groundwater potential and its relation with population in the lower Barpani watershed, Assam, India. *Natural Resources Research, 27*(4), 503–515. https://doi.org/10.1007/s11053-017-9367-y

Alam, W., Zamminsion, K., Gyanendra, Y., Laishram, R. J., & Nesa, N. (2022). Geochemical and multivariate assessment of groundwater resources of churachandpur sub-division of Manipur, India. *Applied Water Science, 12*(5), 1–15. https://doi.org/10.1007/s13201-022-01638-y

Allafta, H., Opp, C., & Patra, S. (2021). Identification of groundwater potential zones using remote sensing and GIS techniques: A case study of the Shatt Al-Arab Basin. *Remote Sensing, 13*(1), 1–28. https://doi.org/10.3390/rs13010112

Bhuyan, M. J., & Deka, N. (2022). Delineation of groundwater potential zones at micro-spatial units of Nagaon district in Assam, India, using GIS-based MCDA and AHP techniques. *Environmental Science and Pollution Research* (Resilient and Sustainable Water Management in Agriculture). https://doi.org/10.1007/s11356-022-24505-4

Borah, H., & Deka, S. (2023). Exploration of potential zones of groundwater in Jamuna watershed, Assam, by applying multi-influencing factor technique. *Journal of the Indian Society of Remote Sensing, 51*(1), 75–91. https://doi.org/10.1007/s12524-022-01625-5

Bordoloi, A., Singh, K. K., & Gaichunglu, G. (2023). Application of analytical hierarchy process and GIS techniques to delineate the groundwater potential zones in and around Jorhat and Majuli areas of Eastern Assam, India. *Modeling Earth Systems and Environment, 9*(2), 1589–1612. https://doi.org/10.1007/s40808-022-01583-4

Das, B., & Pal, S. C. (2019). Combination of GIS and fuzzy-AHP for delineating groundwater recharge potential zones in the critical Goghat-II block of West Bengal, India. *HydroResearch, 2*, 21–30. https://doi.org/10.1016/j.hydres.2019.10.001

Gao, B. (1996). NDWI—A normalized difference water index for remote sensing of vegetation liquid water from space. *Remote Sensing of Environment, 58*(3), 257–266. https://doi.org/10.1016/S0034-4257(96)00067-3

Gholami, V. (2022). Spatial modeling of groundwater depth fluctuations using co-active neuro-fuzzy inference system (CANFIS) and geographic information system (GIS). *Applied Water Science, 12*(3), 1–12. https://doi.org/10.1007/s13201-022-01572-z

Golden, B. L., & Wang, Q. (1989). An alternate measure of consistency. In B. L. Golden, E. A. Wasil, & P. T. Harker (Eds.), *The analytic hierarchy process* (pp. 68–81). Springer. https://doi.org/10.1007/978-3-642-50244-6_5

Gopalakrishnan, G., Zhou, Y., Raj, B., Senapathi, V., & Ramkumar, T. (2018). Remote sensing and GIS based groundwater potential zone mapping in Ariyalur district, Tamil Nadu. *Journal of the Geological Society of India, 92*, 484–490. https://doi.org/10.1007/s12594-018-1046-z

Gyeltshen, S., Kannaujiya, S., Chhetri, I. K., & Chauhan, P. (2023). Delineating groundwater potential zones using an integrated geospatial and geophysical approach in Phuentsholing, Bhutan. *Acta Geophysica, 71*(1), 341–357. https://doi.org/10.1007/s11600-022-00856-x

Mahato, R., Bushi, D., Nimasow, G., Nimasow, O. D., & Joshi, R. C. (2022). AHP and GIS-based delineation of groundwater potential of Papum Pare district of Arunachal Pradesh, India. *Journal of the Geological Society of India, 98*(1), 102–112. https://doi.org/10.1007/s12594-022-1936-y

Maity, B., Mallick, S. K., Das, P., & Rudra, S. (2022). Comparative analysis of groundwater potentiality zone using fuzzy AHP, frequency ratio and Bayesian weights of evidence methods. *Applied Water Science, 12*(4), 1–16. https://doi.org/10.1007/s13201-022-01591-w

Ministry of Water Resources Guwahati. (2013). *Ground water information booklet, North Cachar Hills District, Assam, technical report series: D. Central Ground Water Board (CGWB), North Eastern Region, Ministry of Water Resources, Guwahati, Assam, India.* Available at: cgwb.gov.in/District_Profile/Assam.Cachar

Paksoy, T., Pehlivan, N. Y., & Kahraman, C. (2012). Organizational strategy development in distribution channel management using fuzzy AHP and hierarchical fuzzy TOPSIS. *Expert Systems with Applications, 39*(3), 2822–2841. https://doi.org/10.1016/j.eswa.2011.08.142

Rahaman, M.H., Sajjad, H., Roshani, Masroor, M., Bhuyan, N., & Rehman, S. (2022). Delineating groundwater potential zones using geospatial techniques and fuzzy analytical hierarchy process (FAHP) ensemble in the data-scarce region: Evidence from the lower Thoubal river watershed of Manipur, India. *Arabian Journal of Geosciences, 15*(8), 677. https://doi.org/10.1007/s12517-022-09946-y

Saaty, T. L. (1988). What is the analytic hierarchy process? In G. Mitra, H. J. Greenberg, F. A. Lootsma, M. J. Rijkaert, & H. J. Zimmermann (Eds.), *Mathematical models for decision support* (pp. 109–121). Springer. https://doi.org/10.1007/978-3-642-83555-1_5

Shao, Z., Huq, M. E., Cai, B., Altan, O., & Li, Y. (2020). Integrated remote sensing and GIS approach using fuzzy-AHP to delineate and identify groundwater potential zones in semi-arid Shanxi Province, China. *Environmental Modelling and Software, 134*, Article 104868. https://doi.org/10.1016/j.envsoft.2020.104868

Sresto, M. A., Siddika, S., Haque, M. N., & Saroar, M. (2021). Application of fuzzy analytic hierarchy process and geospatial technology to identify groundwater potential zones in north-west region of Bangladesh. *Environmental Challenges, 5*, Article 100214. https://doi.org/10.1016/j.envc.2021.100214

Swain, S., Taloor, A. K., Dhal, L., Sahoo, S., & Al-Ansari, N. (2022). Impact of climate change on groundwater hydrology: A comprehensive review and current status of the Indian hydrogeology. *Applied Water Science, 12*(6), 1–25. https://doi.org/10.1007/s13201-022-01652-0

Talukdar, S., Mallick, J., Sarkar, S. K., Roy, S. K., Islam, A. R. M. T., Praveen, B., Naikoo, M. W., Rahman, A., & Sobnam, M. (2022). Novel hybrid models to enhance the efficiency of groundwater potentiality model. *Applied Water Science, 12*(4), 1–22. https://doi.org/10.1007/s13201-022-01571-0

Tiwari, A., Ahuja, A., Vishwakarma, B. D., & Jain, K. (2019). Groundwater potential zone (GWPZ) for urban development site suitability analysis in Bhopal, India. *Journal of the Indian Society of Remote Sensing, 47*(11), 1793–1815. https://doi.org/10.1007/s12524-019-01027-0

Waste Tyre-Derived Magnetic Pyrochar Composite in Remediating Oxytetracycline Antibiotic from Wastewater

Upakrishnam Shrivats Sanskrit, Abhisek Mondal, and Brajesh Kumar Dubey

1 Introduction

A significant growth in the number of vehicles globally has resulted in a notable issue of managing waste tyres, which poses a significant threat to the environment (Cunliffe & Williams, 1998; Rahman et al., 2022). Waste tyres account for about 2% of the overall waste (Pilkington, 2021). Huge stockpiles of waste tyres accumulate due to their dumping, thus posing serious health and safety issues. This is a global concern in the developed and the developing countries due to the environmental pollution caused by waste tyres if not disposed off appropriately (Mushunje et al., 2018).

Inappropriate disposal of used tyres results in a range of environmental and health issues, including air and soil pollution, breeding of mosquitoes, and risk of fires. The current challenges associated with the management of waste tyres include lack of effective recycling technologies and infrastructure, inadequate regulations, and low public awareness of the issue. Furthermore, the pandemic-induced economic slowdown also caused reduced demand for recycled tyres, exacerbating the issue of waste tyre disposal. To address this problem, collaborative efforts of governments, industry stakeholders, and consumers are required to promote sustainable tyre production, incentivise recycling and reuse, and enforce stricter regulations for disposing off 'end of life' tyres. Therefore, conversion of waste tyres into useful carbonaceous materials (or composite) is favourable for the environment as it can be used for removal of pollutants while bringing circularity in waste management (Kong et al., 2017).

U. S. Sanskrit · A. Mondal · B. K. Dubey (✉)
Department of Civil Engineering, Indian Institute of Technology Kharagpur, Kharagpur, India
e-mail: bkdubey@civil.iitkgp.ac.in

A. Mondal
Department of Infrastructure Engineering, The University of Melbourne, Melbourne, Australia

In the present study, waste tyres are utilised to synthesise a material that is used in waste remediation, thereby creating a circular economy for waste. The aim of this study was to utilise waste tyre-derived pyrochar (TPC) as a carbon support for photoactive nanocomposites, which were used for removing emerging contaminants from synthetic wastewater. By utilising waste tyre-derived pyrochar as a carbon support, this study attempted to contribute to the sustainable management of waste while addressing the issue of emerging contaminants in wastewater. This approach highlights the potential for integrating waste-to-resource strategies and advanced treatment technologies to create a more sustainable and circular economy for waste management.

1.1 Literature Review

One type of waste that is extremely challenging to recycle is used tyres. This is due to their complicated structure, varied raw material composition, and structure of the rubber used to manufacture tyres. Reusing and/or recycling these wastes is crucial to protect nonrenewable resources because the majority of these wastes comprise petroleum-based products. However, owing to the network structure of vulcanised rubber, raw ingredients in waste tyres cannot be used again directly. For regeneration or recovery, they require devulcanisation or a proper degrading procedure. The primary technological basis for manufacturing rubber products is the irreversible vulcanisation reaction that takes place between natural and synthetic diene rubbers (neoprene and polychloroprene) along with sulphur and a variety of auxiliary chemicals (Sienkiewicz et al., 2012). Transverse bonds join elastomer chains, creating the cross-linked structure of rubber. Owing to the irreversible disposition of their formation via vulcanisation, rubber products, unlike thermoplastic materials, are characterised by their elasticity, insolubility, and inability to be fused again. That is why, recycling them takes a significant investment in time and energy

© Capital Publishing Company, New Delhi, India 2025

I. Haq et al. (eds.), *Water and Wastewater*, https://doi.org/10.1007/978-3-031-94253-2_3

Table 1 Typical composition of passenger and truck tyres (Idris et al., 2019)

Component	Passenger tyres (wt.%)	Truck tyres (wt.%)	Comments
Rubber	47	45	A variety of natural and synthetic rubbers are employed in the manufacturing of rubber products, such as natural rubber (polyisoprene), styrene-butadiene rubber, nitrile rubber, chloroprene rubber, and polybutadiene rubber
Carbon black	21.5	22	These rubbers can be strengthened and made more abrasion-resistant through the addition of steel belts and cord as reinforcement
Metal	16.5	21.5	Steel belts and cord for strength
Textile	5.5	–	Used for reinforcement
Zinc oxide	1	2	To control the vulcanisation process and enhance the physical properties of rubber, stearic acid is frequently used
Sulphur	1	1	Cross-linking of polymer chains within rubber is achieved by the addition of agents, such as sulphur, which hardens the rubber and helps to prevent excessive deformation at elevated temperatures
Additives	7.5	5	Materials such as clay or silica can be used as partial replacements for carbon black

and depends only on destroying the rubber product through mechanical, thermal, or chemical means; therefore, recovering the basic ingredients used to make rubber products is impossible (White & De, 2007).

A typical composition of a tyre is shown in Table 1. In addition to rubber, which accounts for roughly 70–80% of tyre's bulk, steel belts and textile overlays also contribute to a tyre's final shape and functional characteristics (Pehlken & Müller, 2009). However, the presence of steel belts and textile overlays poses a significant challenge in recycling waste tyres. Following multiple recycling attempts, these waste materials are utilised in a variety of applications, such as packaging materials, paver surfaces, rooftops, and children's playground surfaces (Eldin & Senouci, 1993). Nonetheless, a significant quantity of discarded tyres and other rubber-based wastes are disposed off globally using unfavourable methods.

A lot of interest is shown in using alternative methods for recycling of used tyres; one of such methods uses pyrolysis technology to produce beneficial materials from waste tyres. This process involves subjecting the material to high temperatures to undergo both chemical and physical separation into different molecules in an oxygen-free environment. The limited thermal stability of the material's chemical bonds causes them to break down while exposed to heat.

In pyrolysis, the synthesis of new molecules results from thermal breakdown of waste tyres. This enables to obtain materials, such as oil, char, and gas. This characteristic makes pyrolysis a crucial process for modern industry, because it enables the creation of materials having significantly greater value from common resources and trash. During pyrolysis, charcoal and biochar are produced as well as liquid and non-condensable gases (NCGs), such as H_2, CH_4, C_nH_m, CO, CO_2 and N_2. At a normal pyrolysis temperature of 500 °C, the organic parts of tyres are broken down thermally to recover char, gases, and oil as well as steel.

However, harmful gases produced during the combustion of waste tyres are a hindrance, with dioxins being of particular concern (Richards & Agranovski, 2017). In a fixed bed batch reactor, the components of thermal breakdown of a waste tyre during pyrolysis at an experimental temperature of 400–700 °C include 40.26% oil, 47.88% char, and 11.86% gas (weight%) (Aydin & İlkılıç, 2012).

Antibiotic pollution refers to the release of antibiotics and their metabolites into the environment; this occurs through the disposal of unused medication, human and animal excreta, and improper disposal of medical waste. Antibiotic pollution poses a serious threat to public health and the environment, as it leads to the development of antibiotic-resistant bacteria in addition to other negative environmental impacts. Some basic materials used to remove antibiotics from water include activated carbon, clay minerals, and biological methods, such as bioremediation. Activated carbon absorbs antibiotics onto its surface, while clay minerals remove antibiotics through ion exchange and adsorption mechanisms. The studies conducted to treat water containing antibiotic pollutants by using carbon-supported materials are shown in Table 2.

2 Objective

The overall objective of this research was to synthesise waste tyre-derived carbon-supported Fe_3O_4 composite and utilise it for removing emerging contaminants (e.g. antibiotics) from water. The specific objectives of this study are as follows:

1. Synthesis of waste tyre-derived carbon composites (i.e. TPC and $3Fe_3O_4$@TPC (at mass ratio of 1:3)).

Table 2 Summary of the studies conducted to remove antibiotic pollutants from water using carbon-derived material

Pollutant	Material	Findings	References
Ciprofloxacin (CIP), propranolol (PRO), and clomipramine (CLO)	Magnetic tyre pyrochar (MTC) derived from end-of-life tyres (ELTs)	Maximum adsorption capacities were estimated to be 6.94, 6.6, and 10.55 mg/g for CIP, PRO, and CLO, respectively	Feizi et al. (2020)
Sulfanilamide (SA), sulfamethizole (SMT), and sulfamethoxazole (SMX)	Magnetic-activated carbon composites (MACs) synthesised by Fe sludge and biological sludge	Adsorption capacity of 31.5, 40.6, 52.3, and 70.8 mg/g for SA, SMT and SMX, respectively	Wan et al. (2020)
Tetracycline (TC)	Silicate-hydrochar composite	Adsorption capacities of 183.9–189.8 mg/g for TC	Deng et al. (2020)
Monocyclic aromatic hydrocarbons (toluene)	Hydrochar and pyrochar	Pyrochar toluene: 437.8–640 mg/g, hydrochar toluene: 39.9–724 mg/g	Liu et al. (2021)
Atrazine (ATZ)	Sludge-based pyrochar	ATZ removal efficiency of pyrochar was 48.0 ± 5.1, 47.2 ± 1.3, and 58.3 ± 7.0% at pH 5, 7, and 9, respectively	Zhang et al. (2021)
Ciprofloxacin	Ultraporous polyimide-based hollow carbon nanofiber mat	PP-CNFM-1–1 (pp = polyamide acid/polymethyl methacrylate) has high CIP static adsorption capacity (>700 mg g^{-1}) and fast adsorption rate (<1 h)	Dou et al. (2022)
Ciprofloxacin	Sludge pyrochar	SPA-N-1100 (SPA prepared at 1100 °C under N_2 atmosphere) possessed a maximum CIP adsorption capacity of 10.42 mg/g, which was much higher than that of SPA-O-1025 (SPA prepared at 1025 °C under air atmosphere) (1.80 mg/g)	Li et al. (2021)

2. Evaluation of physico-chemical characteristics of all synthesised materials.
3. Evaluation of antibiotics (i.e., oxytetracycline (OTC)) removal and reusability capacity of synthesised materials.

3 Materials and Methods

3.1 Materials

Waste tyres (obtained from a local cycle repairing shop at the Indian Institute of Technology Kharagpur), iron(II) sulphate heptahydrate ($FeSO_4{\cdot}7H_2O$), Fe(III) chloride ($FeCl_3$), 25% ammonium hydroxide (NH_4OH), and ethanol (C_2H_5OH) were used as received from the supplier without further purification.

3.2 Synthesis of Materials

The TPC–Fe_3O_4 composite was synthesised in a stepwise manner. First, the tyre pyrochar was synthesised, followed by pyrochar–Fe_3O_4 composite synthesis.

3.2.1 Synthesis of TPC

Tyre waste was cut into smaller pieces using a scissor and grinded into fine particles using liquid nitrogen by making them brittle. These particles were then passed through a sieve shaker. Mass fraction passing through the 1.18-mm sieve and retained on 600 μm were collected, and used to feed in tube furnace at 600 °C for 3 h at a heating rate of 6 °C per min. The detailed steps are shown in Fig. 1.

3.2.2 Synthesis of 3Fe_3O_4@TPC

1 g of TPC was taken in a 100-mL beaker containing 50-mL milli-Q water and ultrasonicated for 5 min, followed by magnetic stirring at 550 rpm for 5 min. Then, in a separate beaker, 1.2 g of $FeSO_4$.7H2O and 1.4 g of $FeCl_3$ were dissolved in 50 mL of milli-Q water by magnetic stirring at 550 rpm for 5 min, followed by 1 min ultrasonication. The contents of both beakers were mixed, keeping the final theoretical mass ratio of TPC to Fe_3O_4 as 1:3, and magnetic stirring was done for 1 h at 550 rpm. After that, NH_4OH solution (25% strength) was added dropwise to the above mixture at 60 °C till pH of 10 was achieved. The final dispersion was kept for another 1 h at 80 °C at a rotation of 550 rpm. Black precipitate was observed in the beaker and was magnetically

Fig. 1 Steps in the synthesis of pyrochar from tyre waste using pyrolysis

separated using neodymium magnet. The black precipitate was vacuum-dried overnight in an oven at 60 °C and crushed into a fine powder. The steps are shown in Fig. 2.

3.3 Characterisation of Synthesised Materials

Synthesised materials, that is TPC and $3Fe_3O_4$@TPC, were characterised using scanning electron microscopy with energy dispersive X-ray spectroscopy (SEM-EDAX), X-ray diffraction (XRD), value stream mapping (VSM) analysis, and Raman spectroscopy to analyse quantitively and qualitatively the physico-chemical properties (i.e. size, surface morphology, crystallinity, oxidation of the elements present, and magnetic property) of materials.

3.4 Batch-Scale Experiments

Laboratory scale batch experiments (Fig. 3) with synthesised materials were conducted for removing antibiotics (e.g., oxytetracycline) from aqueous phase. Briefly, 100 mL of antibiotic solution (10 mg L^{-1}) was taken in an Erlenmeyer flask, followed by the addition of carbon composite dose (0.5 g L^{-1}), and stirred at 150 rpm for 2 h. Samples were collected at regular intervals, centrifuged for 10 min at 3500 rpm. The supernatant was analysed for residual antibiotic using UV–Visible spectrophotometer.

4 Results and Discussion

4.1 Characterisation of Synthesised Materials

4.1.1 X-Ray Diffraction (XRD)

The XRD spectra of synthesised composites are shown in Fig. 4. Peaks at around 2θ value of 26° and 29° on TPC spectra are graphite peaks showing the presence of graphitic carbon in TPC. Peaks for $3Fe_3O_4$@TPC are around 2θ value of 35°, representing the presence of Fe_3O_4 in the composite.

4.1.2 Magnetic Property of Synthesised Materials

The VSM analysis of composites are shown in Fig. 5. The TPC curve shows no magnetic behaviour. The zero loss in hysteresis loop of the composite shows that the synthesised magnetic material is soft in nature.

4.1.3 Raman Spectroscopy

The Raman characterisation test result of synthesised TPC and $3Fe_3O_4$@TPC composites are shown in Fig. 6. The intense peaks are on the TPC curve showing the presence of graphite. This peak vanishes as more Fe_3O_4 is impregnated.

4.1.4 Scanning Electron Microscopy (SEM)

The present study unveils important insights into the SEM results of TPC and $3Fe_3O_4$@TPC (Fig. 7), which highlight

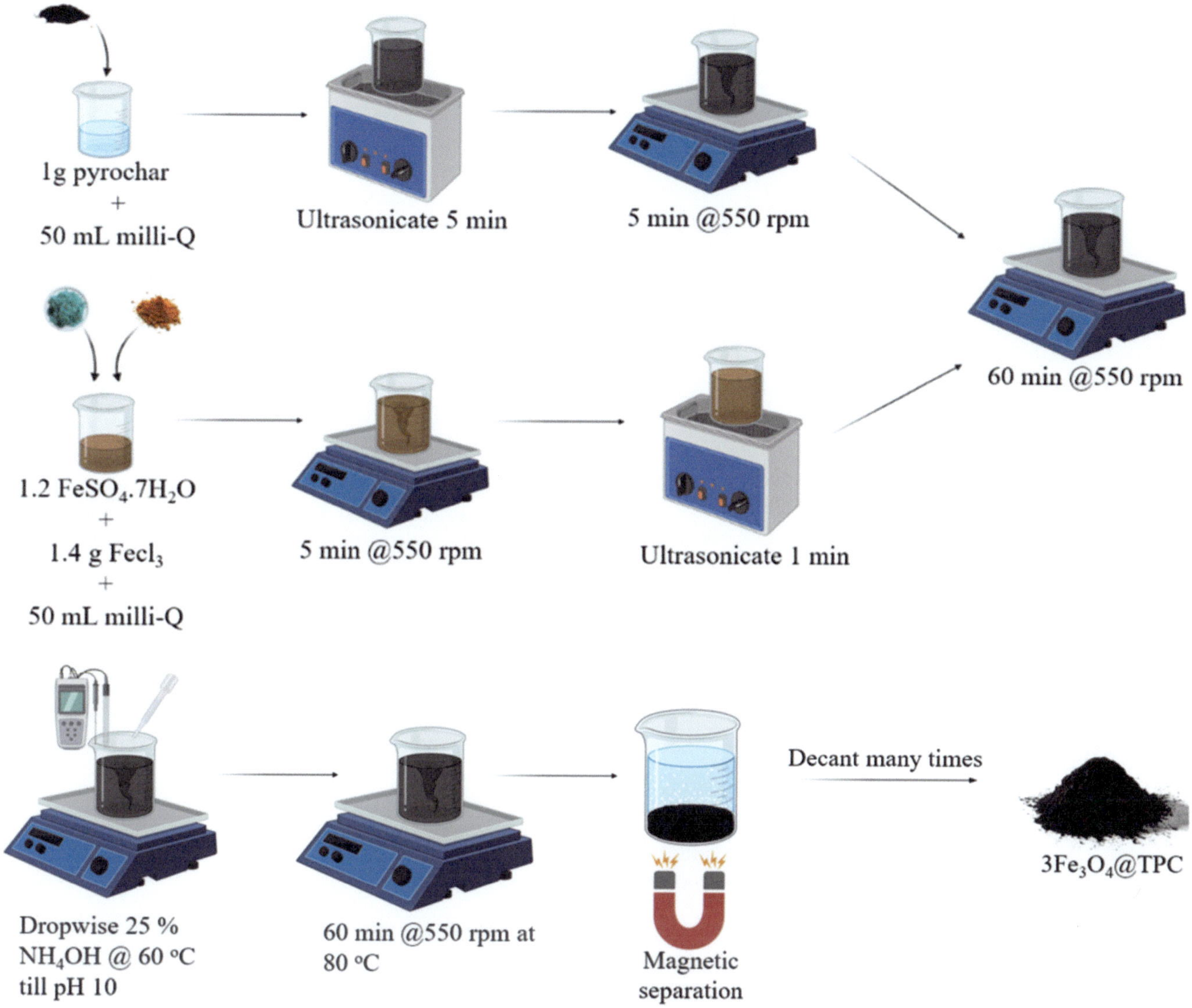

Fig. 2 Stepwise laboratory synthesis of $3Fe_3O_4$@TPC

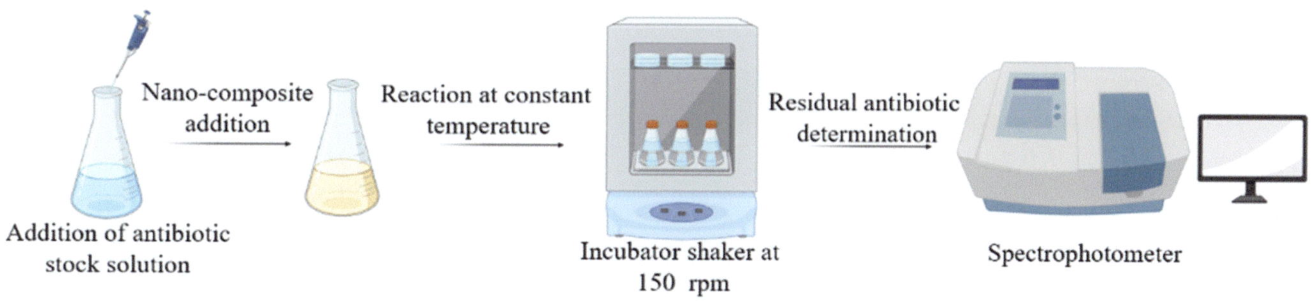

Fig. 3 Schematic for batch-scale experiments

their potential for efficient removal. Notably, our analysis reveals that most of the Fe_3O_4 particles are attached to TPC, resulting in a unique configuration that might enhance the overall material's removal efficiency. These findings offer significant contributions to the development of innovative materials for wastewater treatment, emphasising the importance of material design for achieving optimal performance. Elemental mapping from EDAX is shown in Fig. 8. The elemental analysis shows the distribution of all elements in all synthesised materials. It is used to analyse the presence of

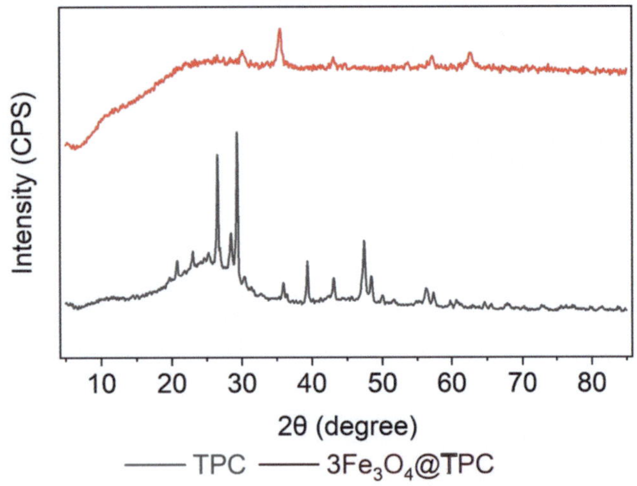

Fig. 4 XRD analysis of TPC and $3Fe_3O_4$@TPC. Set as TPC–$3Fe_3O_4$@TPC through the text, including in all figures and tables

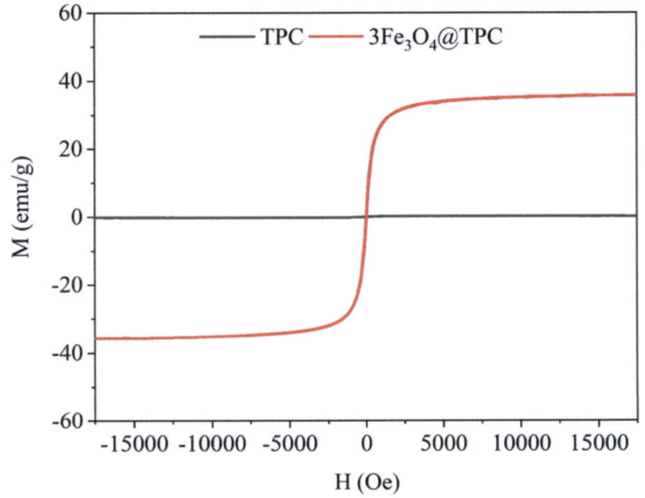

Fig. 5 Magnetic hysteresis loop of TPC and $3Fe_3O_4$@TPC from VSM analysis. Set as "2,000, 2,500, 5,000, 10,000, 15,000 etc." throughout the text

C, O, S, Si, Al, and Fe in synthesised materials to understand elemental distribution.

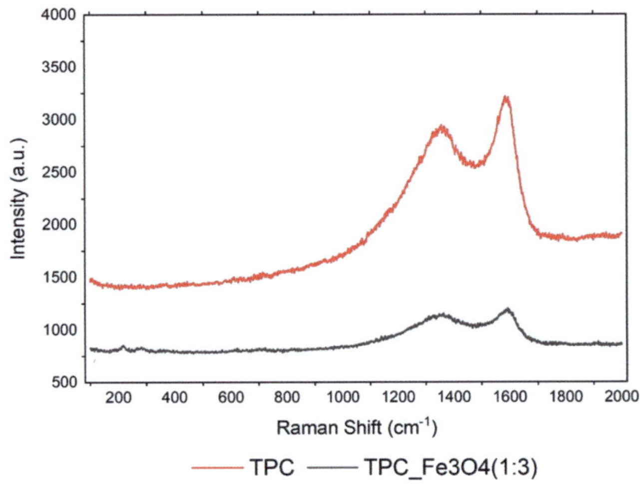

Fig. 6 Raman spectra of TPC and $3Fe_3O_4$@TPC

4.2 Removal Results

4.2.1 Calibration Curve of Oxytetracycline

Graph for calibration of oxytetracycline using UV–Vis spectrophotometry is shown in Fig. 9. Oxytetracycline concentrations of 0, 1, 5, 10, and 15 ppm were used to create calibration curve. Calibration curve was further used to find unknown concentrations during batch reactions.

4.2.2 Removal Efficiency and Reusability Results

Batch-scale experiments were conducted to investigate the removal efficiencies of the oxytetracycline (OTC) using

Fig. 7 SEM images of TPC and $3Fe_3O_4$@TPC

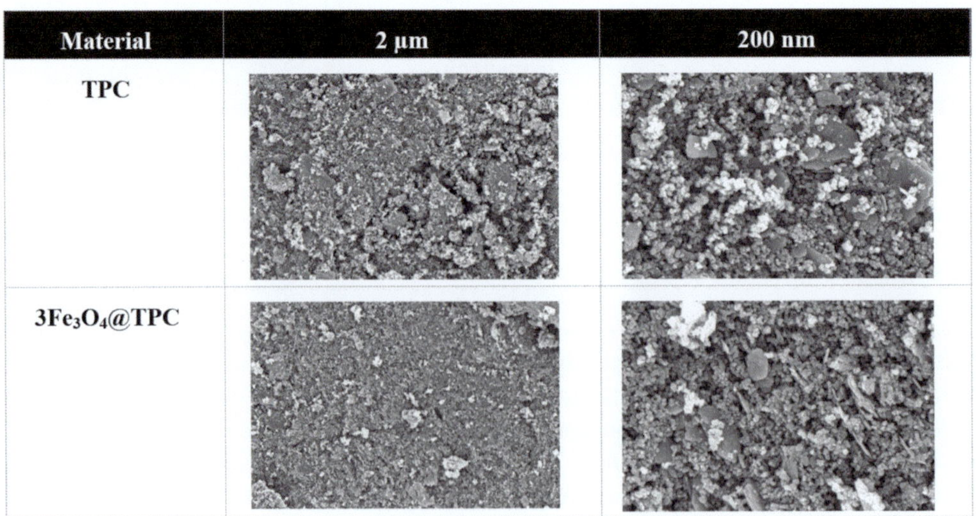

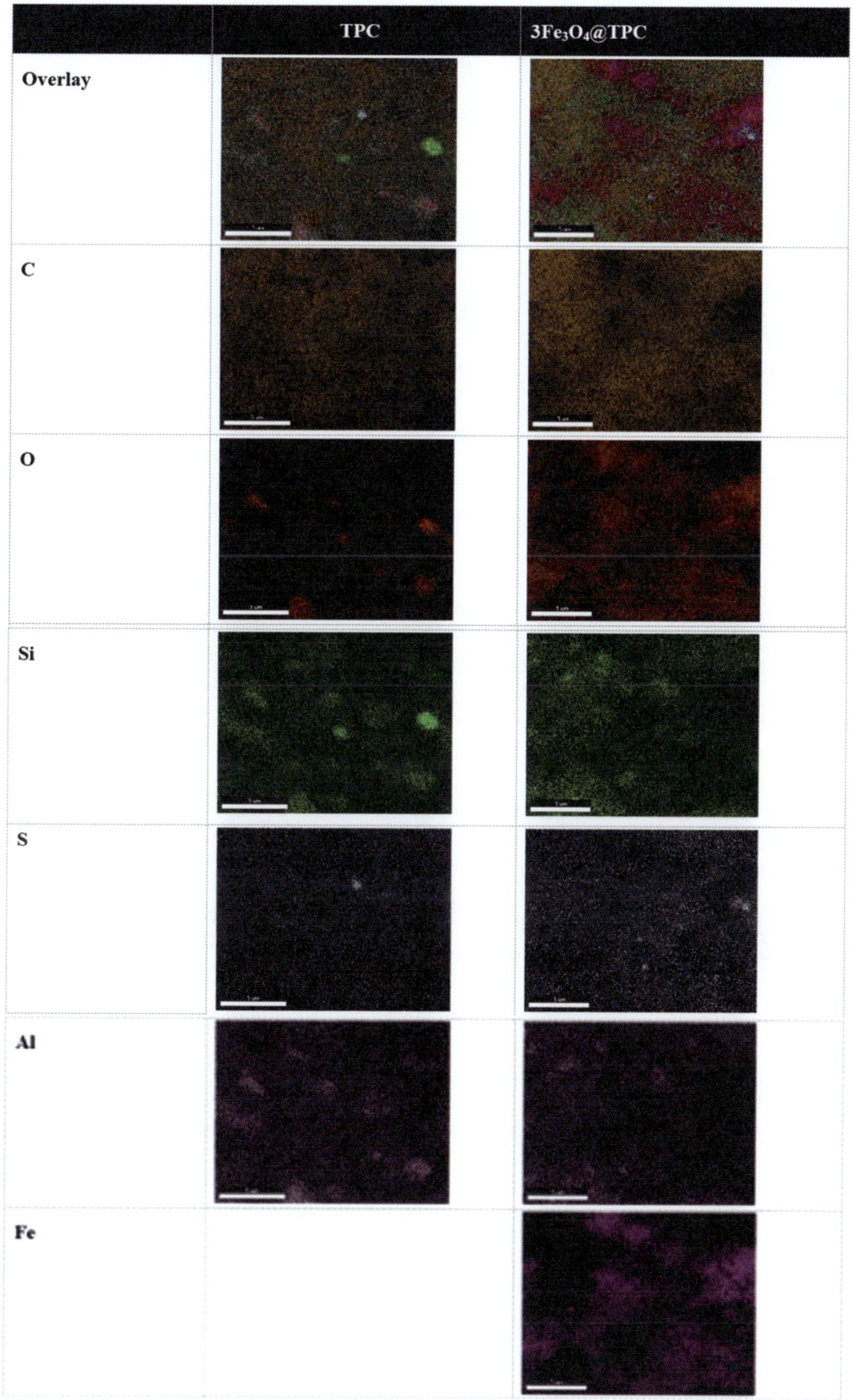

Fig. 8 Elemental overlay of each synthesised material, i.e. TPC and $3Fe_3O_4$@TPC

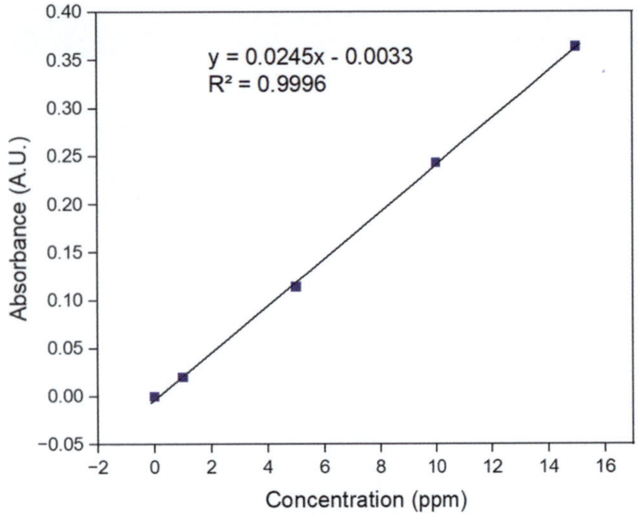

Fig. 9 Calibration curve for oxytetracycline (OTC) using UV–Vis spectrophotometer

two different materials. The results of the experiments are presented in Fig. 10a, where the removal of OTC is reported as follows: TPC (19.70%) < $3Fe_3O_4$@TPC (72.94%). The results indicate that the addition of Fe_3O_4 in TPC contributes to the improved removal of OTC. These findings highlight the potential of newly developed materials for the effective removal of antibiotics from aqueous environments.

The findings regarding the reusability of material $3Fe_3O_4$@TPC, which were investigated over four cycles, is presented in Fig. 10b. The results indicate a decreasing trend in material's performance over the course of four cycles. However, it can be inferred that the synthesised materials are still suitable for multiple cycles of usage, thereby demonstrating their potential to provide an economically viable solution for the process of removal. These findings suggest that further investigation into the material's properties and behaviour is warranted with the aim of optimising their performance and enhancing their potential for reuse.

5 Conclusions

Carbon-based material was derived from waste tyres through pyrolysis. The material was then doped with Fe_3O_4 in a stepwise manner to form $3Fe_3O_4$@TPC composite. The characterisation test XRD showed that Fe_3O_4 is present in the composite, and hence, the material is magnetic. VSM established the nature of magnetism. SEM demonstrated the relative arrangement of TPC and Fe_3O_4. The removal trend was: TPC (19.70%) < $3Fe_3O_4$@TPC (72.94%). Reusability experiments suggested that the material could be reused for several times, thus establishing the process more economical. TPC provided a highly effective adsorptive surface for pollutants, but its recycling could be challenging. To address this, Fe_3O_4 was impregnated to the material, serving a dual purpose of enabling magnetic separation for pollutant removal. The inclusion of carbon-based material was crucial, as Fe_3O_4 composite alone could not facilitate the recycling process because of nano-sized particles. This approach demonstrated great potential for efficient and effective removal of pollutant from wastewater. The magnetic properties of Fe_3O_4 make it an ideal candidate for easy recovery and its reusability as a composite material. Thus, waste tyre-derived magnetic pyro-char composites showed promise for the removal of antibiotics from wastewater, with Fe_3O_4 facilitating separation/removal. This novel approach could pave the way for developing more effective and sustainable materials for wastewater treatment, thus resulting in a cleaner and healthier environment.

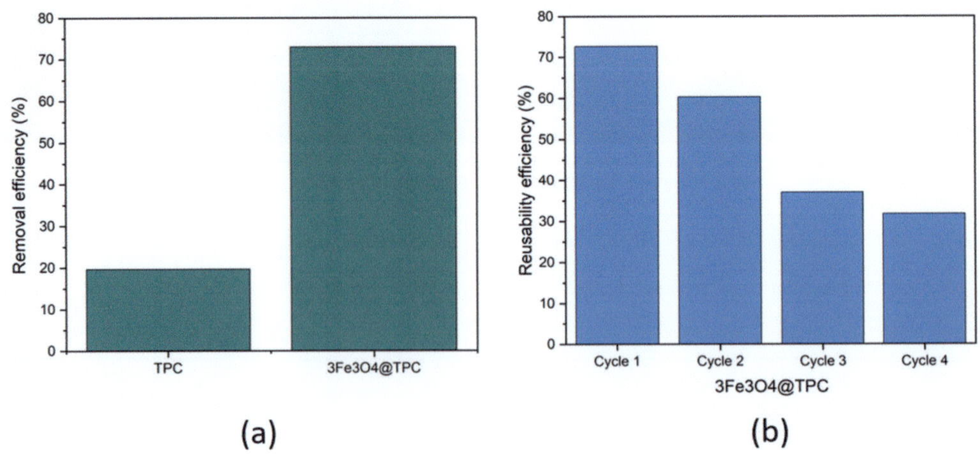

Fig. 10 **a** Removal efficiencies of synthesised materials; **b** the trend of removal efficiency in reusability test for $3Fe_3O_4$@TPC for four cycles

References

Aydın, H., & İlkılıç, C. (2012). Optimization of fuel production from waste vehicle tires by pyrolysis and resembling to diesel fuel by various desulfurization methods. *Fuel, 102*, 605–612. https://doi.org/10.1016/j.fuel.2012.06.067

Cunliffe, A. M., & Williams, P. T. (1998). Composition of oils derived from the batch pyrolysis of tyres. *Journal of Analytical and Applied Pyrolysis, 44*(2), 131–152. https://doi.org/10.1016/s0165-2370(97)00085-5

Deng, J., Li, X., Wei, X., Liu, Y., Liang, J., Song, B., & Huang, W. (2020). Hybrid silicate-hydrochar composite for highly efficient removal of heavy metal and antibiotics: Coadsorption and mechanism. *Chemical Engineering Journal, 387*, Article 124097. https://doi.org/10.1016/j.cej.2020.124097

Dou, S., Ke, X. X., Zhong, L. B., Fan, J. J., Chen, J. P., & Zheng, Y. M. (2022). Novel ultraporous polyimide-based hollow carbon nanofiber mat: Its polymer-blend electrospinning preparation strategy and efficient dynamic adsorption for ciprofloxacin removal. *Separation and Purification Technology, 295*, Article 121341. https://doi.org/10.1016/j.seppur.2022.121341

Eldin, N. N., & Senouci, A. B. (1993). Rubber-tire particles as concrete aggregate. *Journal of Materials in Civil Engineering, 5*(4), 478–496. https://doi.org/10.1061/(asce)0899-1561(1993)5:4(478)

Feizi, F., Reguyal, F., Antoniou, N., Zabaniotou, A., & Sarmah, A. K. (2020). Environmental remediation in circular economy: End of life tyre magnetic pyrochars for adsorptive removal of pharmaceuticals from aqueous solution. *Science of the Total Environment, 739*, Article 139855. https://doi.org/10.1016/j.scitotenv.2020.139855

Idris, R., Chong, C. T., & Ani, F. N. (2019). Microwave-induced pyrolysis of waste truck tyres with carbonaceous susceptor for the production of diesel-like fuel. *Journal of the Energy Institute, 92*(6), 1831–1841. https://doi.org/10.1016/j.joei.2018.11.009

Kong, X., Liu, Y., Pi, J., Li, W., Liao, Q., & Shang, J. (2017). Low-cost magnetic herbal biochar: Characterization and application for antibiotic removal. *Environmental Science and Pollution Research, 24*, 6679–6687. https://doi.org/10.1007/s11356-017-8376-z

Li, J., Pan, L., Yu, G., Li, C., Xie, S., & Wang, Y. (2021). Synthesis of an easily recyclable and safe adsorbent from sludge pyrochar for ciprofloxacin adsorption. *Environmental Research, 192*, Article 110258. https://doi.org/10.1016/j.envres.2020.110258

Liu, Z., Wang, Z., Chen, H., Cai, T., & Liu, Z. (2021). Hydrochar and pyrochar for sorption of pollutants in wastewater and exhaust gas: A critical review. *Environmental Pollution, 268*, Article 115910. https://doi.org/10.1016/j.envpol.2020.115910

Mushunje, K., Otieno, M., & Ballim, Y. (2018). A review of waste tyre rubber as an alternative concrete constituent material. In *MATEC web of conferences* (Vol. 199, p. 11003). EDP Sciences. https://doi.org/10.1051/matecconf/201819911003

Pehlken, A., & Müller, D. H. (2009). Using information of the separation process of recycling scrap tires for process modelling. *Resources, Conservation and Recycling, 54*(2), 140–148. https://doi.org/10.1016/j.resconrec.2009.07.008

Pilkington, B. (2021). *Tackling the global tyre waste problem with pretred*. Editorial, AZoCleantech.com. Available at: https://www.azocleantech.com/article.aspx?ArticleID=1227. Accessed on 27 Oct 2022

Rahman, M. M., Yu, Y., & Wu, H. (2022). Valorisation of waste tyre via pyrolysis: Advances and Perspectives. *Energy & Fuels, 36*(20), 12429–12474. https://doi.org/10.1021/acs.energyfuels.2c02053

Richards, G., & Agranovski, I. E. (2017). Dioxin-like PCB emissions from cement kilns during the use of alternative fuels. *Journal of Hazardous Materials, 323*, 698–709. https://doi.org/10.1016/j.jhazmat.2016.10.040

Sienkiewicz, M., Kucinska-Lipka, J., Janik, H., & Balas, A. (2012). Progress in used tyres management in the European Union: A review. *Waste Management, 32*(10), 1742–1751. https://doi.org/10.1016/j.wasman.2012.05.010

Wan, J., Ding, J., Tan, W., Gao, Y., Sun, S., & He, C. (2020). Magnetic-activated carbon composites derived from iron sludge and biological sludge for sulfonamide antibiotic removal. *Environmental Science and Pollution Research, 27*, 13436–13446. https://doi.org/10.1007/s11356-020-07940-z

White, R., & De, S. (2001). *Rubber technologist's handbook* (Vol. 1). iSmithers Rapra Publishing, Shropshire, UK.

Zhang, L., Wang, K., Yu, L., Luo, K., Guo, S., Chen, H., & Xue, G. (2021). Why does sludge-based hydochar activate peroxydisulfate to remove atrazine more efficiently than pyrochar? *Applied Catalysis b: Environmental, 299*, Article 120663. https://doi.org/10.1016/j.apcatb.2021.120663

Effective Magnetic Loading on Waste Cigarette Filter-Derived Hydrochar for Aqueous Phase Removal of Antibiotics

Eric Gamit, Abhisek Monda, and Brajesh Kumar Dubey

1 Introduction

Cigarettes and its residue, that is cigarette filters/butts, is a major risk factor, such as infections and diseases in humans, other living organisms as well as the environment. Many stray animals eat littered cigarette filters/butts presuming these to be a food, and these leftovers affect the environment because of their slow rate of biodegradation. Owing to their prevalence, toxicity, and longevity, used cigarette butts are a significant and problematic form of waste (CantÃº, 2021). Therefore, disposal of cigarette filters/butts has become one of the major problems nowadays towards which we can't be ignorant. Therefore, conversion of waste cigarette filters/butts to a useful carbonaceous material (or composite) is beneficial for removal of pollutants from the environment. In the process, waste cigarette filters/butts are used to synthesise a material to be used in remediation of waste, thus creating a circular economy for management of waste. This chapter aimed at utilising waste filters/butts-derived hydrochar as carbon support for nanocomposites in removing rising contaminants from synthetic wastewater.

Different methods to recycle cigarette filters/butts are recognised in literature. Cigarette filters were applied without any pre-processing for producing bricks and asphalt. However, different manufacturing techniques are required for additional applications of cigarette filters, such as production of absorbent materials, vector control, and their usage as a biofilm carrier in wastewater treatment. In order to prevent littering of cigarette filters, researchers provided models and recommendations, including penalties, taxes, and public education. In spite of using cutting-edge techniques for cigarette filter recycling as recommended in earlier studies, cigarette filters haven't perceived much attention when it comes to environmental issues and pollution control in recycling operations (Gómez Escobar et al., 2021; Torkashvand & Farzadkia, 2019).

In terms of creating products, recent research has focused on developing methods to repurpose the debris from cigarette filters. A South Korean research team devised a one-step technology to alter cellulose acetate present in used cigarette filters into a high-performing substance that could be used as an energy storage component in computers, mobile devices, electrical vehicles and wind turbines. Evaluating this high-performing material with commercially available carbon, graphene and carbon nanotubes, it had superior performance. Another team of researchers proposed inserting food-grade acid tablets in cigarette filters. Instead of utilising cellulose triacetate and because cigarette smoke is quite acidic, if the inserted tablets are sufficiently moist, they would produce acid that could speed up filter decomposition in about 2 weeks (Lee et al., 2014; Yadav & Hakkarainem, 2021). Table 1 shows different approaches taken by different research groups in synthesising carbon material from waste cigarette filters.

Different researchers around the globe discovered different approaches to remove contaminants from wastewater. Table 2 shows studies on the utilisation of carbon-derived composites to remove contaminants from wastewater.

Many reports and studies have been published globally based on the removal of oxytetracycline and ciprofloxacin from the wastewater using different mechanisms and products; some of those mechanisms are mentioned in Table 3.

E. Gamit
Department of Mechanical Engineering, Indian Institute of Technology Kharagpur, Kharagpur, India

A. Monda · B. K. Dubey (✉)
Department of Civil Engineering, Indian Institute of Technology Kharagpur, Kharagpur, India
e-mail: bkdubey@civil.iitkgp.ac.in

A. Monda
Department of Infrastructure Engineering, The University of Melbourne, Melbourne, Australia

© Capital Publishing Company, New Delhi, India 2025
I. Haq et al. (eds.), *Water and Wastewater*, https://doi.org/10.1007/978-3-031-94253-2_4

Table 1 Use of waste cigarette filters to produce carbon materials

Carbon type	Production method	Qualitative characteristics	Activation agent	References
Activated carbon	Heat in vacuum medium for 3 h at 230 °C	Micro-porous-activated carbon produced had a diameter of 50μm and a length of 10μm	KOH	Hamzah and Umar (2017)
Charred carbon	Direct pyrolysis with 900 C for 3 h at a heating rate of 5 °C/min and in the presence of a constant nitrogen flow of 100 cm^3/min	Activated carbon has a micropore and mesoporous surfaces of 453 m^2/g and 144 m^2/g, respectively	HNO_3	Masoudi Soltani et al. (2015)
Porous carbon	Single-stage pyrolysis at 900 °C for 1 h with a heating rate of 5 °C/min	–	–	Yazdi et al. (2012)

Table 2 An overview of studies on the use of carbon-derived materials to treat antibiotic contaminants

Pollutant	Removal by	Findings	Comments	References
Antibiotic tetracycline	Corn stover as eco-friendly and renewable agro-waste/ biomass as adsorbent	At pH = 7.4, maximum monolayer adsorption capacity (132.9mg/g) was achieved	Achieved high removal capability for antibiotic tetracycline from aqueous media	Haghighat et al. (2020)
Nitroimidazole antibiotics from water	Firstly, porous carbon was derived from pyrolysis of melamine-loaded MAF-6	–	By using ethanol washing, the carbon could be reused with little performance loss	Sarker et al. (2019)
Tetracycline	Prepared a CO_2-derived porous carbon (EC) as adsorbent	EC had a high surface area (1064.8 m^2/g) and comparatively low Zeta potential (2.89). EC had a magnificent absorbability (393.5mg/g) and remarkable regeneration ability	By reduction treatment, tetracycline adsorption on EC was enhanced	Yang et al. (2021)
Tetracycline	From naturally available human hair, KOH-activated carbon was prepared	High surface area (1505.11 m^2/g) and microporosity (68.34%) were obtained. At 50 C, maximum monolayer adsorption capacity of 210.18 mg/g was achieved	For removal of tetracycline, kinetics and isotherm studies were performed	Ahmed et al. (2017)
Carbon type	Production method	Qualitative characteristics	Activation agent	References
Monocyclic aromatic hydrocarbons (Toluene)	Hydrochar and pyrochar	Pyrochar: toluene removal −437.8–640 mg/g Hydrochar: toluene removal −39.9–724 mg/g	Adsorption mechanisms involved electrostatic attraction, ion exchange, precipitation, π–π interaction (mostly occurs in aqueous phase), Van der Waals force, H-bond, and oxidation	Wei et al. (2012)

Table 3 Different removal mechanisms of oxytetracycline and ciprofloxacin from wastewater

Sl. no	Antibiotic	Removal mechanism	Removal efficiency	References
1	Oxytetracycline	AGS by rapid adsorption and slow biodegradation removed oxytetracycline	The degradation proportion in total removal amount of oxytetracycline was 72%	Wang et al. (2019)
2	Oxytetracycline	Removal of one of the antibiotic named oxytetracycline hydrochloride (OTC) by Zn/Fe LDH	OTC removal proportion was 77.23% by Zn/Fe LDH	Zaher et al. (2020)
3	Ciprofloxacin	0.5Fe@CS-T7 exhibited promising performance for peroxydisulfate activation to degrade ciprofloxacin	–	Fan et al., 2022
4	Ciprofloxacin	Synthesis of water-enhanced graphene hydrogel adsorbent. Graphene hydrogel has excellent adsorption properties for ciprofloxacin	–	Yu et al. (2019)

Note LDH: layered double hydroxide; AGS: aerobic granular sludge

The primary objective of this research was to synthesise Fe_3O_4-doped carbon material from used cigarette filters/butts and utilise the same to remove antibiotics from water.

The specific objectives of this study are as follows:

- Synthesis of composites: CFHC, CFHC–Fe_3O_4 (1:1), CFHC–Fe_3O_4 (1:3), and CFHC–Fe_3O_4 (1:5). (The ratio given at this point is: w(hydrochar):w(Fe_3O_4).)
- Evaluation of physicochemical properties of synthesised materials.
- Assessment of antibiotic (i.e. oxytetracycline/ciprofloxacin) removal capacity of synthesised materials from synthetic wastewater.

2 Materials and Methods

2.1 Materials

Cigarette filters were collected from trash at a shop that sells cigarettes near Indian Institute of Technology Kharagpur. Iron(II) sulphate heptahydrate ($FeSO_4{\cdot}7H_2O$), Iron(III) chloride ($FeCl_3$), and 25% ammonium hydroxide (NH_4OH) were utilised as received from the supplier without further purification.

2.2 Material Synthesis

Hydrochar was first synthesised to create CFHC–Fe_3O_4 composite; then Fe_3O_4 was embedded on it at different w/w ratios of 1:1, 1:3, and 1:5.

2.2.1 Synthesis of CFHC

First, cigarette filters were removed from waste cigarettes. Then a beaker containing mixture of cigarette filters and water was stirred for 2 h at 700 rpm and 50 °C. Later the water was discarded, and cigarette filters were rinsed in tap water, drained, and dried in an oven at 60 °C for 2 days. Then dried filters were cut into small pieces using a pair of scissors. After this, in a Teflon reactor, taking cut cigarette filters and deionised water in 1:20 (mass:mass) ratio, hydrochar was synthesised at 220 °C with 6 h residence time in a muffle furnace. Then the mixture was centrifuged for 15 min at around 3500 rpm to separate char from liquid fraction; the char residue was rinsed with milli-Q water and dried overnight in an oven at 60 °C. The char was crunshed into a fine powder using mortar and pestle and stored for further use.

2.2.2 Synthesis of CFHC–Fe_3O_4 Composite

To synthesise CFHC–Fe_3O_4 composite (%weight (1:1)), 0.3 g of CFHC was taken with 50-mL milli-Q water in a beaker (B1), ultrasonicated for 5 min, and stirred for 15 min at

550 rpm. In another beaker (B2), 139.5 mg of $FeCl_3$ and 119.54 mg of $FeSO_4{\cdot}7H_2O$ (weights calculated using total mass of Fe_3O_4 required, i.e. 0.3 g with an [Fe^{2+}]:[Fe^{3+}] ratio of 1:2) were taken in 50 mL of milli-Q water and stirred for 5 min at 550 rpm, followed by 1-min ultrasonication. Contents of beakers B1 and B2 were mixed and stirred for 60 min at 550 rpm. Then 25% NH_4OH was added dropwise at 60 °C till pH of the solution reached 10. The solution was stirred at 550 rpm for an additional 1 h at 80 °C. Finally, the composite was separated magnetically, washed with milli-Q water, and dried overnight in vacuum at 60 °C. The dried material was crushed into a fine powder using mortar and pestle and stored for further use. Similarly, composites of other mass ratios (i.e. 1:3 and 1:5) were synthesised.

2.3 Characterisation of Synthesised Material

All the synthesised composites were characterised using XRD, VSM, Raman spectroscopy and SEM-EDAX. To qualitatively address the existence of Fe(II) and Fe(III) in the synthesised composites, XRD analysis was conducted. The magnetic characteristics of composites were measured by VSM analysis.

2.4 Batch-Scale Experiments

Synthesised materials were implemented in lab-scale batch experiments to remove antibiotics (oxytetracycline and ciprofloxacin). A nanocomposite dose was added to 100 mL of antibiotic solution (5–10 mg/L) in an Erlenmeyer flask, and the solution was swirled at 150 rpm for 2 h with exposure to visible light. Samples were collected in a regular interval, and a UV–Vis spectrophotometer was employed to check for any antibiotic residue. Nanocomposites were collected for solid phase analysis after the reaction using magnets.

3 Results and Discussion

3.1 X-Ray Diffraction—Crystal Structure

The XRD spectra of synthesised CFHC and CFHC–Fe_3O_4 with %weight ratios of 1:1, 1:3, and 1:5 are shown in Fig. 1. The peak of 2θ at approximately 26° in CFHC spectra was due to the presence of graphitic carbon obtained from cigarette filters. However, in different cases of CFHC–Fe_3O_4, the peak of 2θ was approximately 36°, and interestingly, it was observed that the peak became taller and sharper with increase in amount of Fe_3O_4.

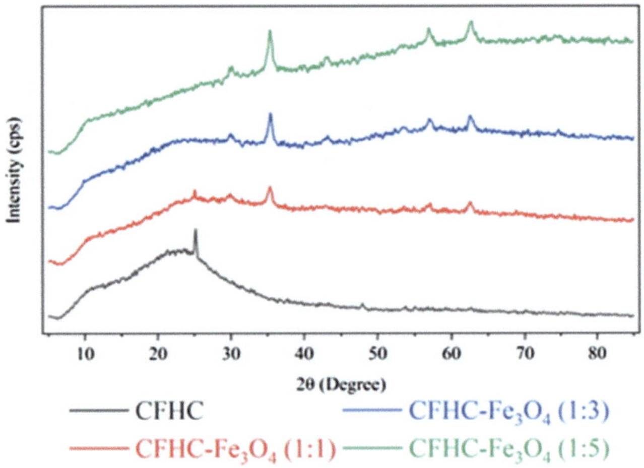

Fig. 1 XRD spectrum of synthesized nanocomposites

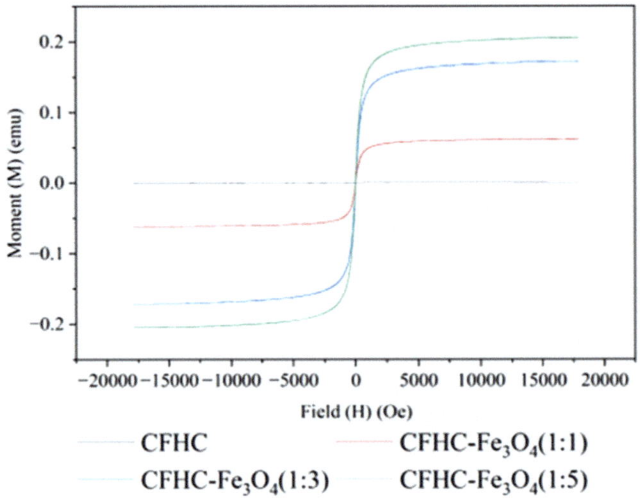

Fig. 2 Magnetic hysteresis of synthesized nanocomposites from VSM analysis

3.2 Vibrational Sample Magnetometer

The hysteresis loop from the VSM analysis of synthesised CFHC and CFHC–Fe_3O_4 composites with %weight ratios of 1:1, 1:3, and 1:5 are shown in. Fig. 2 Zero loss in the area of hysteresis loop suggested that the synthesised magnetic materials were soft in disposition. In addition, it was observed that the value of magnetic saturation increased as the amount of Fe_3O_4 content increased in the material

3.3 Raman Spectrum

The Raman spectrum of synthesised CFHC and CFHC–Fe_3O_4 composites with %weight ratios of (1:3) and (1:5) are shown in Fig. 3. In the case of CFHC, the peak was approximately 100 cm^{-1} (higher) and another approximately

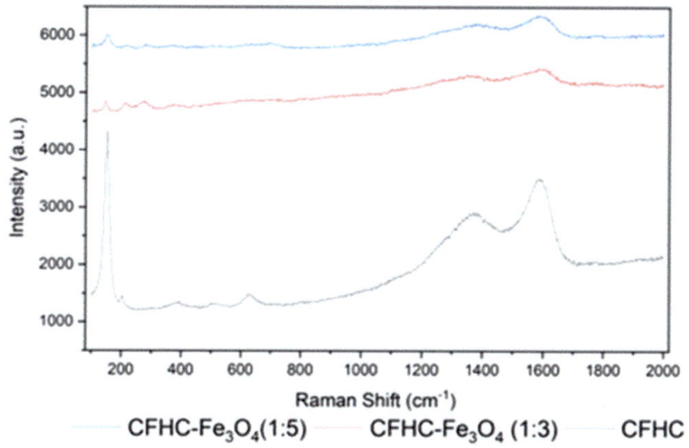

Fig. 3 Raman spectrum

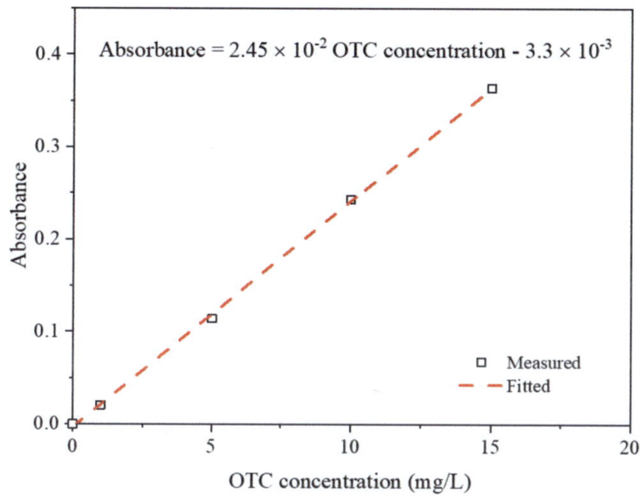

Fig. 4 Calibration curve of Oxytetracycline

1,600 cm^{-1} which suggests grpresence of graphitic carbon. On the other hand, in case of CFHC–Fe_3O_4 composites, the peaks were lower due to the presence of Fe_3O_4 nanoparticles along with CFHC.

3.4 Calibration

For oxytetracycline, data was obtained by using UV–Vis spectrophotometer. First, a calibration curve (Fig. 4) was prepared to establish the relationship between the measured absorbance and the range of oxytetracycline concentrations (Table 4). Measurements are average of duplicate observations.

Table 4 Concentration and respective average absorbance

Concentration (mg/L)	Avg. absorbance
0	0
1	0.02
5	0.11
10	0.24
15	0.36

3.5 Removal Efficiency of Oxytetracycline

Using calibration curve and measured absorbance results, we obtained the efficiencies of oxytetracycline removal from wastewater for different synthesised materials. The efficiencies are provided in Table 5.

3.6 Scanning Electron Microscopy Analysis

In SEM analysis, a powerful investigative method was used to produce intricate, high-magnification photographs of a sample's surface topography using a concentrated electron beam. The obtained SEM images of the materials are shown in Fig. 5.

We observed from the images that increase in the amount of Fe_3O_4 covered more surface area and thus helped to enhance ability to adsorb contaminants from wastewater. However, in the case of CFHC only, we observed that the adsorption was less compared to the adsorption of other materials.

4 Conclusions

X-ray diffraction analysis demonstrated successful synthesis of graphitic hydrochar prepared from waste cigarette filters. Impregnation of Fe_3O_4 nanoparticles on hydrochar surface was evident from XRD analysis showing peaks of approximately 36°, which further turned intense with increasing proportion of Fe_3O_4. Raman spectra suggested presence of graphitic carbon in all hydrochar materials, not visible in XRD analysis, possibly because of the surface coverage of hydrochar particles with Fe_3O_4 nanoparticles. The magnetic analysis of synthesised materials showed ferromagnetic behaviour with increasing magnetic saturation with Fe_3O_4 percentage. Overall, the preliminary characterisation of synthesised materials showed that all materials were magnetic in nature with the presence of graphitic carbon, required

Table 5 Data obtained for final concentrations and respective efficiencies

Sr. no	Material	Initial concentration (mg/L)	Final concentration (mg/L)	Efficiency (%)
1	CFHC	9.69	9.17	5.31
2	Fe_3O_4 + CFHC (1:1)	9.69	8.80	9.18
3	Fe_3O_4 + CFHC (1:3)	11.90	9.36	21.34
4	Fe_3O_4 + CFHC (1:5)	11.90	7.26	39.0

Notes CFHC: cigarette filter hydrochar; Fe_3O_4: magnetite or ferrous-ferric oxide

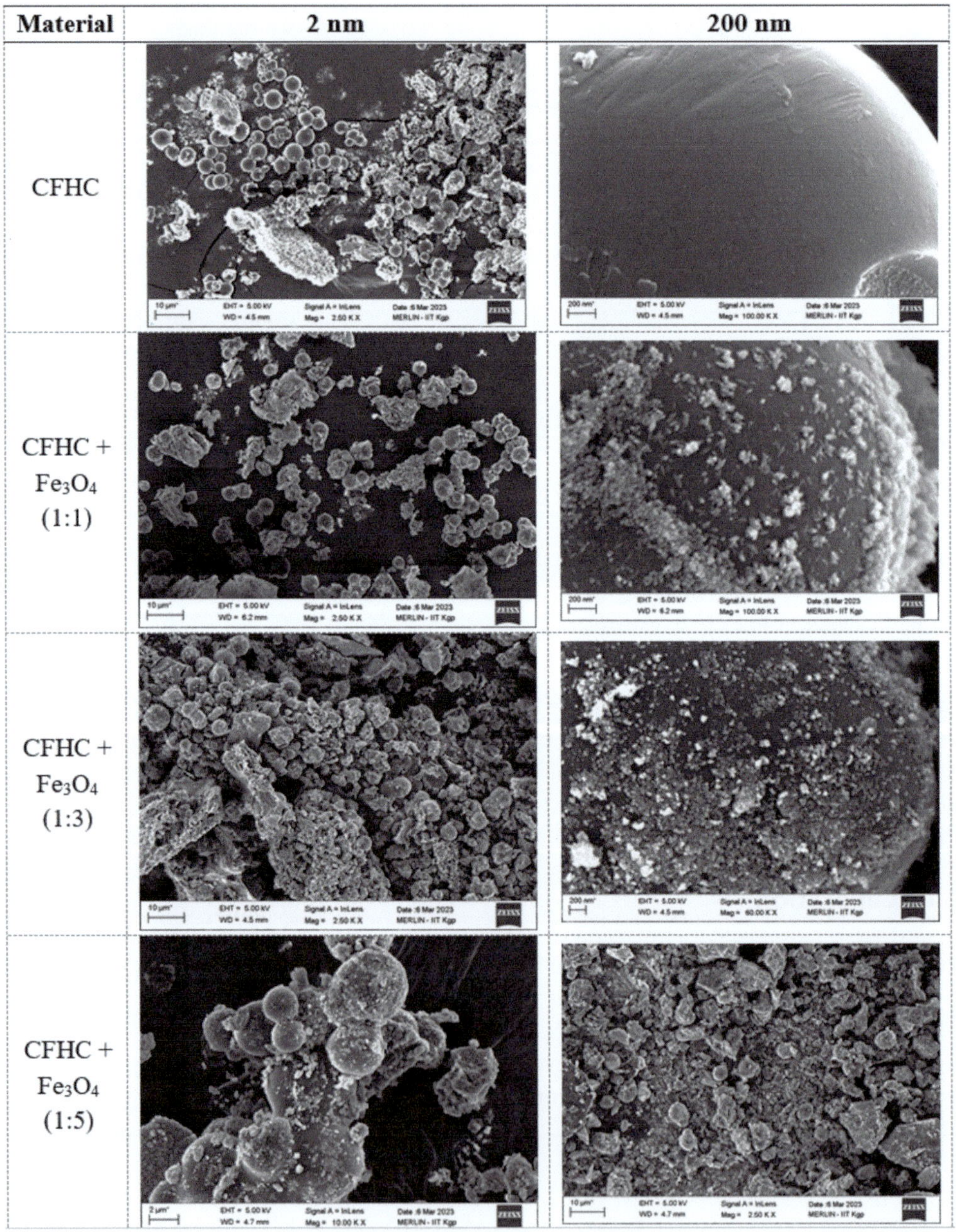

Fig. 5 Scanning electron microscopy (SEM) images of synthesised materials

for the intended remediation application. The SEM analysis established that mere CFHC was not very efficient for removing contaminants from wastewater. Hence, Fe_3O_4 was added to improve the contaminant removal efficiency of wastewater. The equation provided in the graph, obtained from the data, helped to calculate the concentration of an unknown sample with known absorbance to have the idea of suitability of synthesised material for removing oxytetracycline from wastewater.

References

Ahmed, M. J., Islam, M. A., Asif, M., & Hameed, B. H. (2017). Human hair-derived high surface area porous carbon material for the adsorption isotherm and kinetics of tetracycline antibiotics. *Bioresource Technology, 243*, 778–784. https://doi.org/10.1016/j.biortech.2017.06.174

Amal, Z., Taha, M., Farghali, A. A., & Mahmoud, R. K. (2020). Zn/Fe LDH as a clay-like adsorbent for the removal of oxytetracycline from water: Combining experimental results and molecular simulations to understand the removal mechanism. *Environmental Science and Pollution Research, 27*(11), 12256–12269. https://doi.org/10.1007/s11356-020-07750-3. Springer, Berlin, Heidelberg.

CantÃº, A. (2021). New "green" cigarette butts biodegrade within days—and can even sprout into grass. AlterNet.org. Available at: https://www.alternet.org/2014/02/new-green-cigarette-butts-biodegrade-within-days-and-can-even-sprout-grass/. Accessed: 8 Nov 2022.

Fan, X., Lin, Q., Zheng, J., Fu, H., Xu, K., Liu, Y., & He, J. (2022). Peroxydisulfate activation by nano zero-valent iron graphitized carbon materials for ciprofloxacin removal: Effects and mechanism. *Journal of Hazardous Materials, 437*, 129392. https://doi.org/10.1016/j.jhazmat.2022.129392

Gómez Escobar, V., Moreno González, C., Arévalo Caballero, M. J., & Gata Jaramillo, A. M. (2021). Initial conditioning of used cigarette filters for their recycling as acoustical absorber materials. *Materials, 14*(15), 4161. https://doi.org/10.3390/ma14154161

Haghighat, G. A., Saghi, M.H., Anastopoulos, I., Javid, A., Roudbari, A., Talebi, S. S., & Shams, M. (2020). Aminated graphitic carbon derived from corn stover biomass as adsorbent against antibiotic tetracycline: Optimizing the physicochemical parameters. *Journal of Molecular Liquids, 313*: 113523. https://doi.org/10.1016/j.molliq.2020.113523

Hamzah, Y., & Umar, L. (2017). Preparation of creating active carbon from cigarette filter waste using microwave-induced KOH activation. *Journal of Physics: Conference Series, 853*(1), 012027. https://doi.org/10.1088/1742-6596/853/1/012027

Lee, M., Kim, G. P., Song, H. D., Park, S., & Yi, J. (2014). Preparation of energy storage material derived from a used cigarette filter for a supercapacitor electrode. *Nanotechnology, 25*(34), Article 345601. https://doi.org/10.1088/0957-4484/25/34/345601

Masoudi Soltani, S., Kazemi Yazdi, S., Hosseini, S., & Bayestie, I. (2015). Lead removal from aqueous solution using non-modified and nitric acid-modified charred carbon from the pyrolysis of used cigarette filters. *Desalination and Water Treatment, 53*(1), 126–138. https://doi.org/10.1080/19443994.2013.835751

Ni, S., Wang, X., Zhou, G., Yang, F., Wang, J., Wang, Q., & He, D. (2009). Hydrothermal synthesis of Fe_3O_4 nanoparticles and its application in lithium ion battery. *Materials Letters, 63*(30), 2701–2703. https://doi.org/10.1016/j.matlet.2009.09.047

Phumying, S., Labuayai, S., Thomas, C., Amornkitbamrung, V., Swatsitang, E., & Maensiri, S. (2013). Aloe vera plant-extracted solution hydrothermal synthesis and magnetic properties of magnetite (Fe_3O_4) nanoparticles. *Applied Physics A, 111*, 1187–1193. https://doi.org/10.1007/s00339-012-7340-5

Sarker, M., Shin, S., & Jhung, S. H. (2019). Adsorptive removal of nitroimidazole antibiotics from water using porous carbons derived from melamine-loaded MAF-6. *Journal of Hazardous Materials, 378*, Article 120761. https://doi.org/10.1016/j.jhazmat.2019.120761

Torkashvand, J., & Farzadkia, M. (2019). A systematic review on cigarette butt management as a hazardous waste and prevalent litter: Control and recycling. *Environmental Science and Pollution Research, 26*, 11618–11630. https://doi.org/10.1007/s11356-019-04250-x

Wang, X., Shen, J., Kang, J., Zhao, X., & Chen, Z. (2019). Mechanism of oxytetracycline removal by aerobic granular sludge in SBR. *Water Research, 161*, 308–318. https://doi.org/10.1016/j.watres.2019.06.014

Wei, Y., Han, B., Hu, X., Lin, Y., Wang, X., & Deng, X. (2012). Synthesis of Fe_3O_4 nanoparticles and their magnetic properties. *Procedia Engineering, 27*, 632–637. https://doi.org/10.1016/j.proeng.2011.12.498

Yadav, N., & Hakkarainen, M. (2021). Degradable or not? Cellulose acetate as a model for complicated interplay between structure, environment and degradation. *Chemosphere, 265*, Article 128731. https://doi.org/10.1016/j.chemosphere.2020.128731

Yang, J., Dou, Y., Yang, H., & Wang, D. (2021). A novel porous carbon derived from CO_2 for high-efficient tetracycline adsorption: Behavior and mechanism. *Applied Surface Science, 538*, Article 148110. https://doi.org/10.1016/j.apsusc.2020.148110

Yazdi, S. K., Soltani, S. M., & Hosseini, S. (2012). An investigation into the optimum carbonization conditions for the production of porous carbon from a solid waste. *Advanced Materials Research, 587*, 88–92. https://doi.org/10.4028/www.scientific.net/amr.587.88

Yu, F., Sun, Y., Yang, M., & Ma, J. (2019). Adsorption mechanism and effect of moisture contents on ciprofloxacin removal by three-dimensional porous graphene hydrogel. *Journal of Hazardous Materials, 374*, 195–202. https://doi.org/10.1016/j.jhazmat.2019.04.021

Employing Multivariate Statistics as a Tool to Address the Pollution Loadings of an Aquatic Ecosystem

Siddhant Dash, Saswati Ray, and Ajay S. Kalamdhad

1 Introduction

Globally, water bodies have been victims of incessant deterioration in their quality and quantity via different anthropogenic activities. Such has been the scenario that the availability of safe and potable water in the coming decades would be a matter of grave concern, provided the degradation rate continues at the current pace (Lund Schlamovitz & Becker, 2021; Salehi, 2022). From the quality viewpoint, the surface water has been degraded significantly due to untreated municipal and industrial wastewater discharges and leaching from landfills proximate to the surface water body (Chen & Chen, 2021; Collins et al., 2018; Preisner, 2020; Ray et al., 2021a). On the other hand, groundwater has been contaminated primarily due to the percolation of landfill leachates in the absence of appropriate liner systems (Rayet al., 2021b). Furthermore, when it comes to quantity, the groundwater resources are declining rapidly due to their over-dependency for domestic and agricultural requirements (Castellazzi et al., 2018; Dalin et al., 2017; Jain et al., 2021). This has rendered a decrease in base flow, eventually subsiding water levels in surface water bodies. In addition to the declining groundwater, rapid urbanisation and industrial exploitation have contributed to the spatial decline of surface water bodies. Conservation and restoration of water bodies are primarily attributed to quantity and quality deterioration, respectively. While the conservation options are more inclined towards management strategies, restoration options are purely scientific and require scientific reasonings.

In order to have a deeper understanding of the health of a water body, water quality monitoring programmes are essential. These monitoring programmes provide reliable information regarding the spatial and temporal variability of different water quality parameters, thus identifying the problematic parameters and their sources (Dash et al., 2018). However, large monitoring programmes involving numerous locations and sizeable parameters result in complicated and enormous datasets that pose challenges to understand and interpret (Dash et al., 2020a). Therefore, the use of scientific and mathematical tools for ease of interpretation becomes inevitable.

Since the outset of research in this domain, researchers worldwide have adopted indexing as a tool to assess water quality. Water quality indices (WQIs) are mathematical representations of a dataset that produces a numerical value based on the measured water quality of a body (Dash & Kalamdhad, 2021a). For this purpose, certain weights are assigned to different quality parameters, and WQI values are computed based on their observed and standard values. Over the years, significant research has been conducted on water quality indexing. The initial years witnessed surveying techniques being adopted, which are still in practice, for assigning weights to parameters (Anh et al., 2021; Bansah et al., 2018; Kumar & Alappat, 2009; Ngasala et al., 2019; Proulx et al., 2010; Sutadian et al., 2017; Wanda et al., 2014). Herein, specific questionnaires wereprepared and circulated among the general public or experts in different fields. Based on the responses recorded, weights were allocated to various parameters, which eventually led to the estimation of WQI values. However, several limitations in employing the method, such as allowing bias and prejudice because of subjective approach through individual perceptions, and the technique

S. Dash (✉)
Department of Civil Engineering, School of Engineering and Sciences, SRM University-AP, Guntur, Andhra Pradesh, India
e-mail: dash.siddhant93@gmail.com

S. Ray
Department of Civil Engineering, PP Savani University, Dhamdod, Kosamba, Gujarat, India

A. S. Kalamdhad
Department of Civil Engineering, Indian Institute of Technology Guwahati, Guwahati, Assam, India

© Capital Publishing Company, New Delhi, India 2025
I. Haq et al. (eds.), *Water and Wastewater*, https://doi.org/10.1007/978-3-031-94253-2_5

often deemed unscientific, resulted in the search for new scientific methods for developing novel indices (Dash & Kalamdhad, 2021b). Some of these limitations included statistics (Howladar et al., 2018; Mostafaei, 2014; Patil et al., 2020; Singh et al., 2016; Varol & Davraz, 2015), fuzzy logic and probability (Lermontov et al., 2009; Nasiri et al., 2007; Yaseen et al., 2018) and information entropy (Adimalla et al., 2020; Dash & Kalamdhad, 2021c; Islam et al., 2017; Singaraju et al., 2018; Wang et al., 2009). A detailed bibliographic survey of the history of WQIs is presented by Dash and Kalamdhad (2021b).

All the methods adopted are capable of handling massive data effectively without incorporating loss of important information. The most effective approaches use multivariate statistics and information entropy as well as the concept of probability. However, in the case of smaller scale monitoring programmes (i.e. where the frequency of sampling is less), the use of information entropy and the idea of probability becomes inappropriate. This is because the probability values tend to be either 1 or 0, with a decrease in sample size. This provides some degree of certainty to probability estimation, which becomes debatable, as it is evident that natural systems are highly dynamic. Hence, given the subjective unscientific reputation (considering experts' opinions), it became essential that efficacy of the multivariate statistical approach be checked. Currently, the primary tool used by researchers globally for assigning weights through multivariate statistics is principal component analysis (PCA). The entire water quality dataset for a particular body is first consolidated, and then PCA is performed. This results in the principal components that are then used for the computation of WQI values. Nevertheless, no study has still provided efficacy in the application of this method. Therefore, it is also crucial that the present use of assigning weights through multivariate statistics is checked for correctness and reliability.

Considering the following objectives, the present study aims to discuss the current practice, identify the limitations, and propose a novel approach to assign weights through multivariate statistics. The results of this study would provide a new direction towards assessing the water quality of a body by employing multivariate statistics. In addition, substantial evidence for checking the reliability and correctness of the method would be achieved to help design engineers and policymakers in undertaking necessary steps towards restoring water bodies from potential pollution.

2 Materials and Methods

2.1 Study Area

Deepor Beel (Fig. 1) is one of the 42 Ramsar sites in India and one of the three in the country's north-eastern region. Situated in the southern banks of the Brahmaputra River, the wetland acts as an ecological hotspot, providing essential services to the entire biological community as a whole (Dash et al., 2020b). However, it is subjected to rampant exploitation and suffers from an existential crisis. Once, it stretched to a massive 40 km^2 area, but now the effective area of the wetland has been reduced to just over 4 km^2 (Dash & Kalamdhad, 2021c). Significant urban built-up is responsible for the wetland's precarious condition. As shown in Fig. 1, the entire wetland is surrounded by urban built-ups on the three sides, while the southern portion consists of the Rani and Garbhanga forest reserve. In the eastern part lies the Boragaon landfill, where Guwahati city's municipal solid wastes are dumped (Dash et al., 2018). Additionally, the Basistha River, carrying the city's domestic wastewater, confluences with Deepor Beel in this region. Furthermore, in the western front, a surge in the development of numerous small- and large-scale industries has taken place, discharging their wastes directly into the wetland (Dash et al., 2019). The Khanajan channel serves as an outlet to the wetland where it merges with the Brahmaputra River. Therefore, a significant threat is observed to the wetland regarding both space and quality with the current levels of persistent anthropogenic contributions.

2.2 Sample Collection and Laboratory Analyses

Monthly analytical monitoring of the water column was conducted from October 2017 to February 2019. This followed a preliminary inspection of the study area that helped to identify monitoring sites based on their accessibility and proximity to a pollution source. The water samples were collected from 23 monitoring locations at a depth of 30 cm from the top surface of water level. Two high-density *polyethylene* (HDPE) bottles, pre-cleaned and rinsed with deionised water, were used for collection of samples from each site. One bottle was used for physicochemical analyses, while the other was used for heavy metal analyses. A few drops of concentrated nitric acid

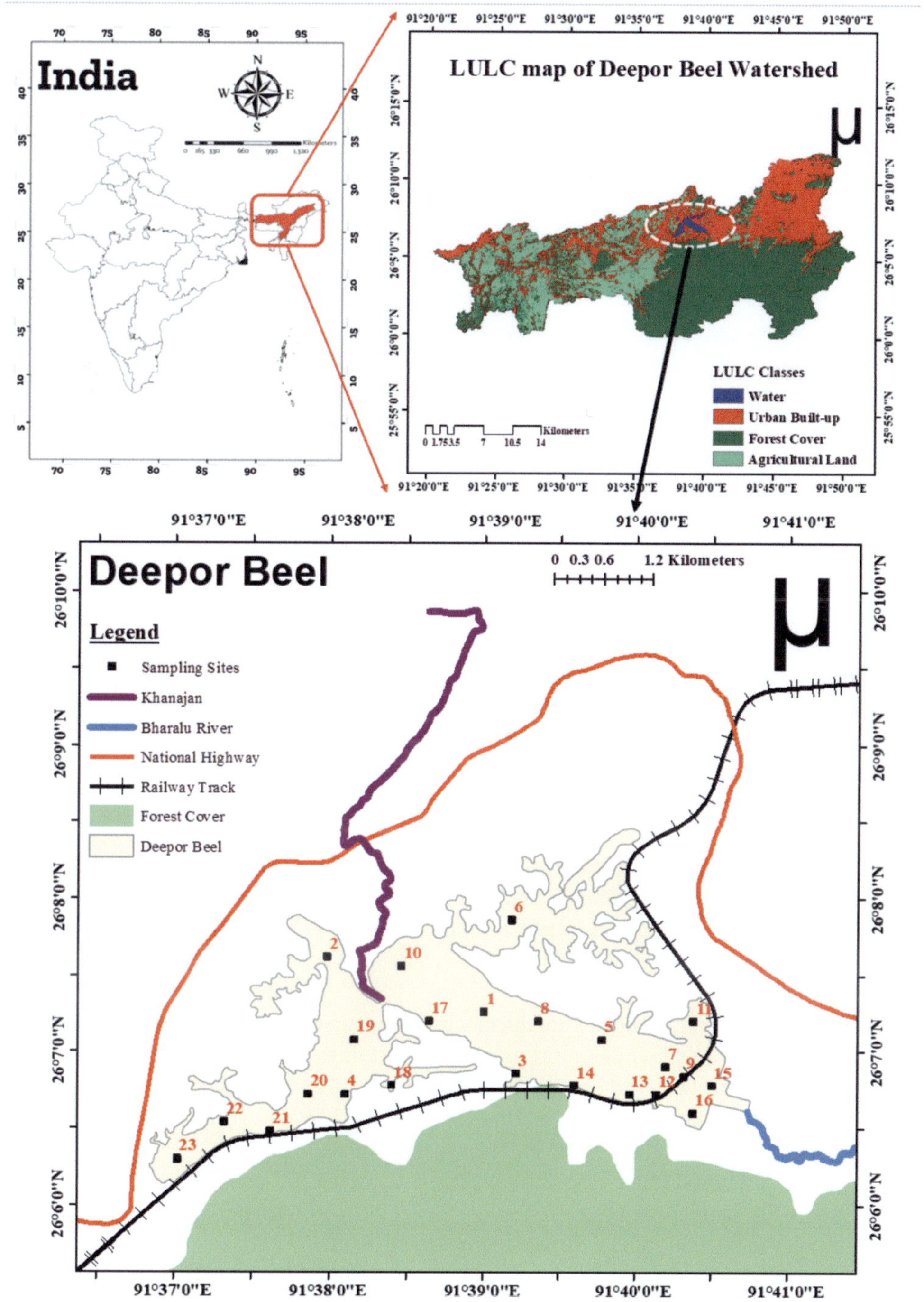

Fig. 1 Study area (Deepor Beel) with land use land cover (LULC) map of its watershed and sampling locations

(HNO_3) were added into the bottle containing water sample for heavy metal analyses until the sample's pH dropped below 2.0. This was done to avoid precipitation in the water sample, because precipitation hinders analytical process and provides erroneous results. All samples were analysed for 24 water quality parameters (Table 2). pH, electrical conductivity (EC) and turbidity (TUR) were measured in situ. For analyses of other parameters, the samples were transported by storing them in an icebox to avoid changes in their properties. The entire process of collection, storage, transportation and analysis of water samples was conducted adhering to the standard methods (American Public Health Association [APHA], 2012).

2.3 PCA–WQI: Its Current Use and Proposed Modification

Multivariate statistical techniques consist of the following four primary tools: PCA, cluster analysis (CA), discriminant analysis (DA) and positive matrix factorisation (PMF) (Dash et al., 2020a). These tools are responsible for determining the overall status of a water body through source identification, classification and apportionment, respectively. In the present practice, the first step of data analysis is PCA. The entire water quality dataset is consolidated in PCA, which is accomplished to obtain component loading values. These loading values are then used to determine WQI. CA is performed to classify sampling sites into different clusters based on their intra-cluster resemblance and inter-cluster variance. However, a major issue exists, which is explained as follows.

It is evident that all monitoring sites behave differently under varying environments depending on pollution sources. For this reason, it is essential to first recognise the sites comprising similar attributes to reveal latent pollution sources influencing their water quality. When conducted on the overall dataset, the PCA allocates weights because of reductions in the dimensionality of dataset without reflecting on the pollution factors of the sites of each cluster. This means that for a study area, the weights of parameters for the locations close to an industrial complex where the primary pollutants are heavy metals are the same as for the sites near a sewage discharge facility where the primary contributors are biological oxygen demand over 5 days (BOD_5), chlorine (Cl^-) and chemical oxygen demand (COD). This argument thus becomes debatable.

On the contrary, exercising hierarchical clustering first to the comprehensive dataset classifies monitoring locations based on their similarities. The PCA then aids in data reduction and recognising pollution sources of each cluster. The weights assigned to each cluster then emphasise the sources responsible for pollution of the respective cluster and not the entire water body in aggregate.

Therefore, the desired modifications in employing multivariate statistics to compute WQI values are explained through the following steps.

Step I: CA of the consolidated water quality dataset

The first step of the proposed methodology is to consolidate all observations into a single dataset. The consolidated dataset is then normalised, considering that all parameters have different units of measurement and they are to be brought under a single unit. The transformation is carried out through z-transformation, after which the clustering process is initiated. This is exercised through Ward's clustering method with clusters varying by a squared Euclidean interval (Wilks, 2011). This results in the categorisation of sampling sites into different clusters, such that each site is first viewed as a single cluster, and then sites having closest resemblances to one another with respect to water quality parameters are separated under the same cluster. The process is repeated until all sites are grouped under a single large cluster.

Step II: Employing PCA on each cluster

Following the CA, the PCA is performed for independent clusters. The sampling adequacy is confirmed using the Keiser–Meyer–Olkin (KMO) criterion (Kaiser, 1960), and the applicability of PCA is authenticated through Bartlett's test of sphericity (Dalal et al., 2010). Principal component (PC) loading values are obtained for all water quality parameters and independent clusters. The PC loading values obtained for each cluster are applicable for all sampling sites under that particular cluster. PCs are rotated using varimax rotation, such that the factors are orthogonal to one another (Tabachnick et al., 2007). Finally, factors with eigenvalues >1 are considered for the determination of loadings.

Step III: Determination of sub-indices/weights

The sub-index values of each parameter and cluster are computed by determining the product of relative eigenvalues and relative loading values. Both relative values were obtained by dividing absolute values by the summation of all values, thus making the overall sum as 1. This applies to all clusters obtained from CA. In case of a single factor, relative eigenvalue is considered 1, and the relative loading value is calculated accordingly.

Step IV: Estimation of PCA–WQI

The final step comprises calculating PCA–WQI values for all clusters, given by the following formula.

$$PCA\text{–}WQI = \sum_{p=1}^{n} \left\{ W_p \times \frac{O_p}{S_p} \right\} \times 100, \qquad (1)$$

where w_p denotes the weights and O_p and S_p represent the observed and standard prescribed values for the p parameter, respectively.

Table 1 Water quality classification for PCA–WQI

Range of WQI values	Category
WQI < 50	Excellent
50 ≤ WQI < 100	Good
100 ≤ WQI < 200	Poor
200 ≤ WQI < 300	Very poor
WQI ≥ 300	Unsuitable for drinking

The present study obtained PCA–WQI values for both current and proposed methodologies for 23 sampling sites and 24 parameters spanning from October 2017 to February 2019. Months of January–March were considered as a pre-monsoon period, April–September as a monsoon season and October–December as a post-monsoon period. In addition to this, a comparative assessment of both methods was carried out to understand differences in both scenarios. For both cases, standard values were considered according to the Bureau of Indian Standards (BIS:10500, 2012). The classification of monitoring sites based on PCA–WQI is shown in Table 1.

2.4 Sensitivity Assay

Every method/model requires assessment for its correctness and reliability. This is essential to test the applicability of the method/model in different scenarios. Sensitivity analysis provides a test for both reliability and correctness of a model. In addition, it provides evidence of the impact of input parameters on the assessed results and other variables used in the model (Gao et al., 2020). Therefore, in the present study, sensitivity analysis was performed on the results obtained by the proposed methodology to determine its usage efficacy. The sensitivity value of each parameter was determined employing the following formula.

$$S_p(\%) = \frac{\left|\left\{\frac{U_p}{N} - \frac{u_p}{n}\right\}\right|}{U_p}, \quad (2)$$

where U_p and u_p indicate PCA–WQI values with and without the pth component (parameter) and N and n are the number of parameters for computing U_p and u_p, respectively.

3 Results and Discussions

3.1 Descriptive Statistics of the Water Quality (WQ) Dataset

Table 2 summarises the descriptive statistical analyses of water quality dataset. Almost all parameters were observed to exceed the permissible limits prescribed by BIS:10, 500 (2012). In addition, slightly higher standard deviations (SD) were observed for dissolved oxygen (DO), turbidity, total alkalinity (TA), total hardness (TH) and total dissolved solids (TDS in ppm). Furthermore, only copper (Cu) exhibited negative skewness, whereas seven parameters (DO, pH, EC, magnesium [Mg^{2+}], cadmium [Cd], iron [Fe] and manganese [Mn]) showed slightly positive skewness (skewness < 1). All other parameters exhibited high positive skewness (i.e. skewness > 1), thereby indicating having a long right-hand tail curve if distributed normally. Finally, 50% of the parameters exhibited a leptokurtic-type distribution (kurtosis > 3), that is, they had a flat curve if distributed normally. In contrast, remaining parameters displayed a plytokurtic-type distribution, that is, they possessed a high peak when the data was normally distributed. For more details, spatiotemporal pattern plots of all water quality parameters are referred to (Dash & Kalamdhad, 2021c).

3.2 PCA–WQI: Current Practice Versus Proposed Methodology

For the current practice of computing PCA–WQI, the entire dataset was compiled, and PCA was conducted. This resulted in six principal components, accounting for more than 87% of cumulative variance (Table 3). PCs with maximum absolute value of each parameter were selected (highlighted in red fonts) to compute sub-indices/weights (Table 4). These weights were employed to determine the PCA–WQI values of individual sampling sites and all three seasons. Figure 2 shows the spatiotemporal variations of PCA–WQI for the current practice method.

It was observed that the pre-monsoon period had the best water quality in the wetland, with the most sites placed in the 'good' category. Exceptions, such as sites proximate to the Boragaon landfill showed 'poor' water quality. Few locations near the industrial zone were on the borderline

Table 2 Descriptive statistics of different water quality parameters used in the study

Parameter	Analytical methodology	Make and model	Minimum	Maximum	Mean	Std. Dev	Skewness	Kurtosis
DO	Winkler's method		0.980	17.532	7.034	2.994	0.231	−0.075
pH	Digital pH meter	Systronics	5.310	8.507	6.725	0.519	0.643	0.755
EC	Digital conductivity meter	VSI-04-Deluxe	0.131	0.583	0.284	0.075	0.791	0.731
Turbidity	Nephelometric turbidimeter	Systronics	0.300	114.500	21.164	20.262	1.654	2.558
TA	APHA titrimetric method		20.062	182.000	84.153	29.833	1.308	1.708
TH	APHA titrimetric method		30.000	150.000	66.489	21.182	1.757	3.401
BOD_5	5-day BOD test		4.500	98.600	20.267	14.277	2.285	7.725
TDS	Oven drying at 103–105 °C		30.685	874.667	218.697	143.348	1.674	3.841
F^-			0.002	5.710	0.407	0.443	9.475	107.621
Cl^-			3.767	123.368	20.560	12.764	4.147	24.495
NO_3^-	Ion chromatograph (IC)	792 Basic IC Metrohm	0.001	30.106	1.650	3.383	5.312	34.970
PO_4^{3-}			0.003	4.993	0.921	1.129	1.607	1.566
SO_4^{2-}			2.408	110.812	16.383	14.297	3.086	12.778
Na^+			0.023	48.200	8.187	9.907	1.376	1.126
K^+	Flame photometer	Systronics	0.031	34.100	4.409	4.588	1.582	5.609
Ca^{2+}			4.215	184.000	56.197	28.020	1.390	3.175
Mg^{2+}			1.000	9.975	4.854	2.292	0.158	−1.246
Cr			0.002	0.466	0.044	0.037	4.961	48.577
Cd			0.001	0.016	0.006	0.003	0.111	−0.346
Fe	Atomic absorption spectroscopy (AAS)	Varian Spectra 55B	0.049	1.606	0.553	0.281	0.113	−0.851
Mn			0.047	0.870	0.308	0.145	0.675	0.442
Cu			0.045	0.980	0.469	0.229	−0.120	−1.288
Pb			0.003	0.088	0.022	0.015	2.572	6.941
NH_4^-	Semi-automated colorimetry method		0.198	5.569	1.194	0.696	2.116	7.237

Table 3 Varimax-rotated component matrix for different water quality parameters (PCA was carried out on the entire dataset)

Eigenvalue	10.877	3.087	2.691	1.569	1.515	1.196
Variance (%)	45.322	12.864	11.212	6.537	6.312	4.984
Cumulative variance (%)	45.322	58.186	69.398	75.935	82.247	87.230
Principal component	1	2	3	4	5	6
DO	−0.836	−0.272	0.184	0.291	0.065	0.243
pH	−0.825	−0.264	0.188	0.326	0.142	0.225
EC	0.865	0.421	−0.093	−0.157	−0.038	−0.063
Turbidity	0.516	0.021	−0.32	0.007	0.254	0.047
TA	−0.825	−0.346	0.22	0.268	0.081	0.202
TH	−0.029	−0.134	0.901	0.008	0.002	−0.006
BOD_5	0.843	0.469	−0.086	−0.042	−0.057	−0.002
TDS	0.838	0.477	−0.126	−0.089	−0.063	−0.026
F^-	−0.043	0.002	−0.034	−0.025	−0.911	0.116
Cl^-	0.274	0.761	−0.098	0.03	−0.263	0.197
NO_3^-	−0.161	−0.12	0.376	0.23	0.101	−0.759
PO_4^{3-}	0.222	0.28	−0.079	0.781	0.036	−0.118
SO_4^{2-}	−0.08	0.782	−0.179	0.115	0.237	0.162
Na^+	0.21	−0.255	0.621	−0.346	0.226	−0.429
K^+	0.192	0.362	0.343	−0.545	0.524	0.223
Ca^{2+}	−0.128	−0.154	0.911	−0.057	0.04	−0.164
Mg^{2+}	0.959	0.072	0.013	0.183	−0.013	0.113
Cr	0.919	0.137	−0.062	0.18	0.083	0.174
Cd	0.788	−0.339	0.22	−0.168	0.287	0.07
Fe	0.963	−0.091	0.028	0.061	0.02	0.109
Mn	0.958	0.078	0.036	0.192	0.047	0.067
Cu	0.949	−0.146	0.09	0.134	0.078	0.041
Pb	0.95	0.17	0.049	0.09	0.123	−0.015
NH_4^-	0.317	0.689	−0.194	0.112	0.058	−0.211

Table 4 Weights of different variables (PCA was carried out on the entire dataset)

PC	Eigenvalue	Relative eigenvalue	Variable	Loading value	Relative loading value	Weight
1	10.877	0.520	DO	0.836	0.069	0.036
			pH	0.825	0.069	0.036
			EC	0.865	0.072	0.037
			Turbidity	0.516	0.043	0.022
			TA	0.825	0.069	0.036
			BOD_5	0.843	0.070	0.036
			TDS	0.838	0.070	0.036
			Mg^{2+}	0.959	0.080	0.041
			Cr	0.919	0.076	0.040
			Cd	0.788	0.065	0.034
			Fe	0.963	0.080	0.042
			Mn	0.958	0.080	0.041
			Cu	0.949	0.079	0.041
			Pb	0.950	0.079	0.041
			Total	12.034		
2	3.087	0.147	Cl^-	0.761	0.341	0.050
			SO_4^{2-}	0.782	0.350	0.052
			NH_4^-	0.689	0.309	0.046
			Total	2.232		
3	2.691	0.129	TH	0.901	0.370	0.048
			Na^+	0.621	0.255	0.033
			Ca^{2+}	0.911	0.374	0.048
			Total	2.433		
4	1.569	0.075	PO_4^{3-}	0.781	0.589	0.044
			K^+	0.545	0.411	0.031
			Total	1.326		
5	1.515	0.072	F^-	0.911	1.000	0.072
			Total	0.911		
6	1.196	0.057	NO_3^-	0.759	1.000	0.057
			Total	0.759		

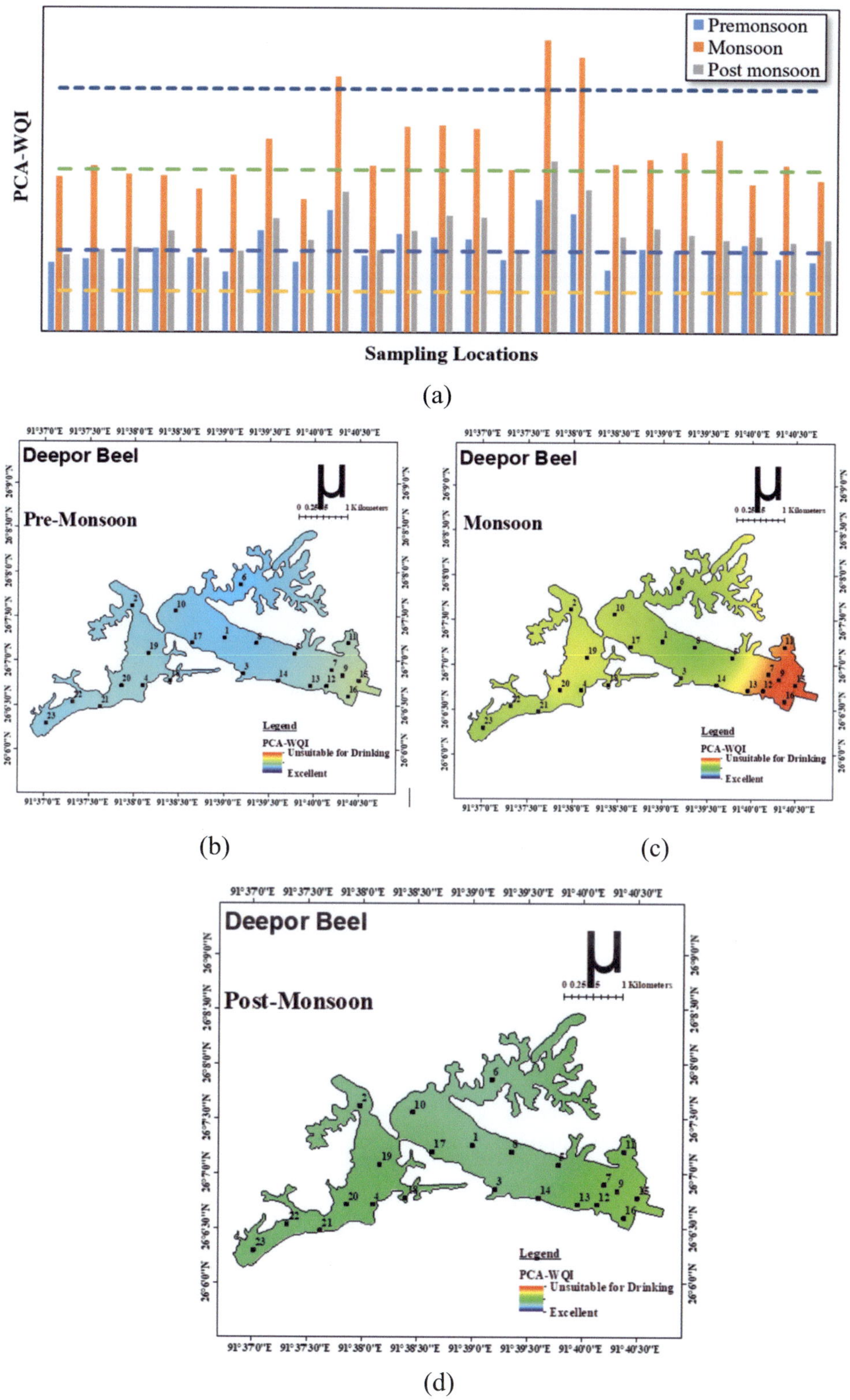

Fig. 2 Spatiotemporal variation of PCA–WQI computed through the current method of practice

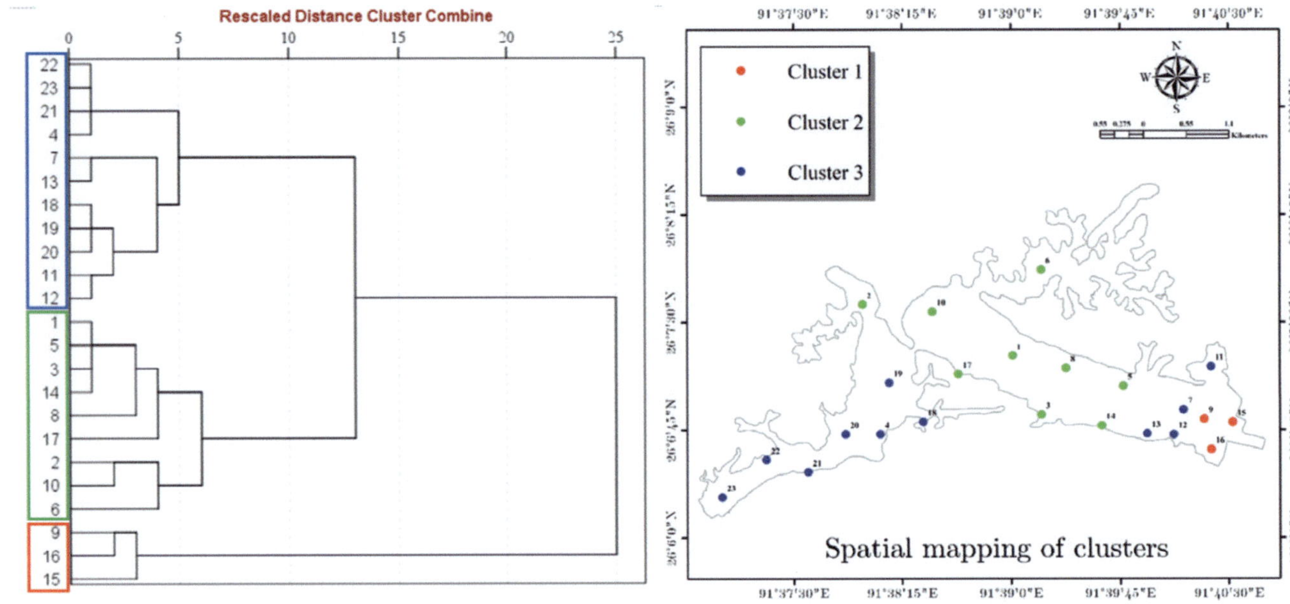

(a) (b)

Fig. 3 Result of cluster analysis through **a** a dendrogram and **b** its spatial extent

of 'good' and 'poor' water quality. During the monsoon season, water quality in the wetland was worst and most sites decelerated to 'poor' and 'very poor' categories, especially for the locations closer to the landfill. This was primarily attributed to leaching from the landfill during this period. In addition, sites in the industrial zone are placed mainly in the 'very poor' category because of excessive discharge of heavy metals from industries. Sites during the post-monsoon period displayed a uniform variation of water quality, with maximum areas placed in 'very poor' category. Few locations indicated 'good' water quality, while only a handful displayed 'very good' water quality.

To compute PCA–WQI for all locations employing the proposed methodology, the water quality dataset of Deepor Beel was first subjected to hierarchical clustering. This resulted in a dendrogram (Fig. 3a), which showed three prominent clusters, indicated by red, green and blue boxes. The same was plotted as a map, as shown in Fig. 3b. Clearly, the sites closest to the landfill showed similar characteristics, followed by the areas in the middle of the landfill and finally the locations in industrial zone. Furthermore, locations 7, 11, 12 and 13 demonstrated similar traits as in cluster 3. Thus, it could be stated that cluster 1 comprised sites with maximum pollution load, followed by cluster 3 having sites with moderate pollution load and finally cluster 2 having sites with minimum pollution load. Sites 7, 11, 12 and 13 transitioned between clusters 1 and 2. Once the three clusters were demarcated, the water quality of each cluster was separated, and PCA was conducted separately. The results of PCA for all three clusters are displayed in Table 5 (clusters a–c).

PCA of cluster 1 resulted in two components, whereas clusters 2 and 3 resulted in five components. The highest component value for each parameter was identified (as highlighted in red fonts) and the corresponding weights were assigned (Table 6, clusters a–c). The weights were then used to compute PCA–WQI values for the sites corresponding to the respective cluster. Figure 4 indicates results of the spatiotemporal variation of PCA–WQI employing the proposed approach. It was observed that both pre-monsoon and post-monsoon periods showed near-similar water quality, with most sites placed under 'poor' category. However, the post-monsoon period had slightly poorer quality than the pre-monsoon period. The same was reflected from spatiotemporal maps (Fig. 4b and d). The monsoon season reflected the worst scenario, wherein the water quality was 'unsuitable for drinking' for sites 15 and 16. These sites were closest to the landfill and the outlet of the Basistha River that discharges city's municipal wastewater into the wetland. In addition, most sites were placed in 'very poor' or 'poor' category. This was primarily attributed to excessive leaching, heavy metal discharges from the landfill and nutrient contamination from the Basistha River. In addition, owing to surface water runoff, the middle section of the wetland was subjected to severe turbidity in water column, rendering the water quality as 'poor'.

A comparative assessment of both results indicated significant differences, especially during the monsoon season, where PCA–WQI of the current usage suggested water quality as 'good' at many sites. In contrast, the results obtained by PCA–WQI based on the proposed methodology indicated 'substantially poorer' water quality. In addition, during the

Table 5 Varimax-rotated component matrix for different water quality parameters (PCA was carried out after HCA on individual clusters)

(a) *Cluster 1*		
Eigenvalue	14.627	9.373
Variance (%)	60.945	39.055
Cumulative variance (%)	60.945	100
Principal component	1	2
DO	−0.994	0.111
pH	−0.952	−0.306
EC	0.932	−0.363
Turbidity	−0.195	−0.981
TA	−0.932	0.363
TH	0.261	−0.965
BOD_5	0.996	0.087
TDS	0.994	−0.112
F^-	0.321	0.947
Cl^-	0.819	−0.574
NO_3^-	−0.215	0.977
PO_4^{3-}	−0.818	−0.576
SO_4^{2-}	0.946	0.323
Na^+	−0.363	0.932
K^+	0.429	0.903
Ca^{2+}	0.958	0.286
Mg^{2+}	0.995	−0.100
Cr	0.975	0.224
Cd	0.975	−0.22
Fe	0.892	0.451
Mn	0.985	0.171
Cu	0.999	0.039
Pb	0.966	0.259
NH_4^-	−0.293	−0.956

(b) *Cluster 2*					
Eigenvalue	12.892	4.204	2.445	2.089	1.256
Variance (%)	53.715	17.516	10.186	8.702	5.234
Cumulative variance (%)	53.715	71.231	81.418	90.12	95.354
Principal component	1	2	3	4	5
Do	−0.899	−0.318	0.185	−0.018	−0.081
pH	−0.864	−0.294	0.326	0.169	−0.002
EC	0.912	0.382	−0.066	0.072	0.103
Turbidity	0.743	0.555	0.114	0.176	0.041
TA	−0.85	−0.332	0.281	0.007	−0.044
TH	0.111	−0.214	0.943	−0.042	−0.144
BOD_5	0.877	0.457	−0.073	0.044	0.1
TDS	0.871	0.446	−0.11	0.031	0.099
F^-	0.015	−0.032	−0.057	−0.992	0.028

(continued)

Table 5 (continued)

Cl^-	0.155	−0.145	−0.388	−0.576	0.658
NO_3^-	−0.55	0.187	0.785	0.164	−0.056
PO_4^{3-}	0.816	0.384	0.378	0.005	0.109
SO_4^{2-}	−0.108	−0.389	−0.14	0.272	0.819
Na^+	−0.011	0.143	0.648	0.412	−0.142
K^+	0.201	−0.111	0.064	0.767	0.501
Ca^{2+}	−0.238	−0.199	0.937	−0.036	−0.109
Mg^{2+}	0.361	0.922	−0.087	−0.046	−0.082
Cr	0.411	0.89	−0.149	−0.033	−0.091
Cd	0.608	0.786	0.014	0.057	0.062
Fe	0.274	0.924	−0.158	−0.058	−0.153
Mn	0.414	0.908	0.022	0.04	−0.008
Cu	0.261	0.961	0.018	0.047	−0.06
Pb	0.565	0.807	0.058	0.084	0.068
NH_4^-	0.453	0.233	−0.229	−0.006	0.776
(c) *Cluster 3*					
Eigenvalue	11.928	4.064	2.607	1.414	1.355
Variance (%)	49.702	16.933	10.864	5.893	5.644
Cumulative variance (%)	49.702	66.634	77.499	83.392	89.036
Principal component	1	2	3	4	5
DO	−0.974	0.05	−0.065	0.087	0.096
pH	−0.827	−0.024	−0.363	0.275	−0.078
EC	0.978	−0.006	−0.053	−0.031	−0.039
Turbidity	−0.35	−0.644	0.096	−0.264	0.602
TA	−0.976	0.012	−0.05	−0.086	0.103
TH	−0.289	0.541	0.268	−0.06	−0.646
BOD_5	0.961	0.011	0.053	−0.129	−0.056
TDS	0.95	0.000	0.008	−0.132	−0.131
F^-	0.184	−0.197	−0.312	−0.496	−0.31
Cl^-	−0.253	0.445	0.255	0.287	−0.049
NO_3^-	0.265	0.087	0.907	0.249	−0.007
PO_4^{3-}	0.201	−0.271	−0.015	0.914	0.024
SO_4^{2-}	0.038	0.007	−0.077	0.179	0.767
Na^+	−0.072	0.469	0.786	0.027	-0.314
K^+	−0.599	0.75	−0.11	−0.14	0.197
Ca^{2+}	0.05	0.876	0.339	−0.101	−0.136
Mg^{2+}	0.972	−0.102	−0.029	0.175	0.03
Cr	0.971	−0.085	−0.046	0.138	0.099
Cd	0.939	−0.01	−0.096	0.143	0.191
Fe	0.966	−0.142	0.066	0.123	0.066
Mn	0.977	−0.121	0.02	0.116	0.079
Cu	0.912	−0.321	0.117	0.049	−0.045
Pb	0.961	−0.006	−0.033	0.174	0.099
NH_4^-	0.163	0.42	0.307	0.631	0.161

Table 6 Weights of different variables (PCA was carried out after HCA on individual clusters)

PC	Eigenvalue	Relative eigenvalue	Variable	Loading value	Relative loading value	Weight
(a) *Cluster 1*						
1	14.627	0.609	DO	0.994	0.062	0.038
			pH	0.952	0.059	0.036
			EC	0.932	0.058	0.035
			TA	0.932	0.058	0.035
			BOD_5	0.996	0.062	0.038
			TDS	0.994	0.062	0.038
			Cl^-	0.819	0.051	0.031
			PO_4^{3-}	0.818	0.051	0.031
			SO_4^{2-}	0.946	0.059	0.036
			Ca^{2+}	0.958	0.059	0.036
			Mg^{2+}	0.995	0.062	0.038
			Cr	0.975	0.060	0.037
			Cd	0.975	0.060	0.037
			Fe	0.892	0.055	0.034
			Mn	0.985	0.061	0.037
			Cu	0.999	0.062	0.038
			Pb	0.966	0.060	0.037
			Total	16.128		
2	9.373	0.391	Turbidity	0.981	0.147	0.058
			TH	0.965	0.145	0.057
			F^-	0.947	0.142	0.056
			NO_3^-	0.977	0.147	0.057
			Na^+	0.932	0.140	0.055
			K^+	0.903	0.136	0.053
			NH_4^-	0.956	0.144	0.056
			Total	6.661		
(b) *Cluster 2*						
1	12.892	0.563	DO	0.899	0.132	0.074
			pH	0.864	0.126	0.071
			EC	0.912	0.133	0.075
			Turbidity	0.743	0.109	0.061
			TA	0.850	0.124	0.070
			BOD_5	0.877	0.128	0.072
			TDS	0.871	0.127	0.072
			PO_4^{3-}	0.816	0.119	0.067
			Total	6.832		
2	4.204	0.184	Mg^{2+}	0.922	0.149	0.027
			Cr	0.890	0.144	0.026
			Cd	0.786	0.127	0.023
			Fe	0.924	0.149	0.027
			Mn	0.908	0.146	0.027
			Cu	0.961	0.155	0.028
			Pb	0.807	0.130	0.024

(continued)

Table 6 (continued)

PC	Eigenvalue	Relative eigenvalue	Variable	Loading value	Relative loading value	Weight
			Total	6.198		
3	2.445	0.107	TH	0.943	0.285	0.030
			NO_3^-	0.785	0.237	0.025
			Na^+	0.648	0.196	0.021
			Ca^{2+}	0.937	0.283	0.030
			Total	3.313		
4	2.089	0.091	F^-	0.992	0.564	0.051
			K^+	0.767	0.436	0.040
			Total	1.759		
5	1.256	0.055	Cl^-	0.658	0.292	0.016
			SO_4^{2-}	0.819	0.364	0.020
			NH_4^-	0.776	0.344	0.019
			Total	2.253		
(c) *Cluster 3*						
1	11.928	0.558	DO	0.974	0.079	0.044
			pH	0.827	0.067	0.037
			EC	0.978	0.079	0.044
			TA	0.976	0.079	0.044
			BOD_5	0.961	0.078	0.043
			TDS	0.950	0.077	0.043
			Mg^{2+}	0.972	0.079	0.044
			Cr	0.971	0.079	0.044
			Cd	0.939	0.076	0.042
			Fe	0.966	0.078	0.044
			Mn	0.977	0.079	0.044
			Cu	0.912	0.074	0.041
			Pb	0.961	0.078	0.043
			Total	12.364		
2	4.064	0.190	Turbidity	0.644	0.237	0.045
			Cl^-	0.445	0.164	0.031
			K^+	0.750	0.276	0.053
			Ca^{2+}	0.876	0.323	0.061
			Total	2.715		
3	2.607	0.122	NO_3^-	0.907	0.536	0.065
			Na^+	0.786	0.464	0.057
			Total	1.693		
4	1.414	0.066	F^-	0.496	0.243	0.016
			PO_4^{3-}	0.914	0.448	0.030
			NH_4^-	0.631	0.309	0.020
			Total	2.041		
5	1.355	0.063	TH	0.646	0.457	0.029
			SO_4^{2-}	0.767	0.543	0.034
			Total	1.413		

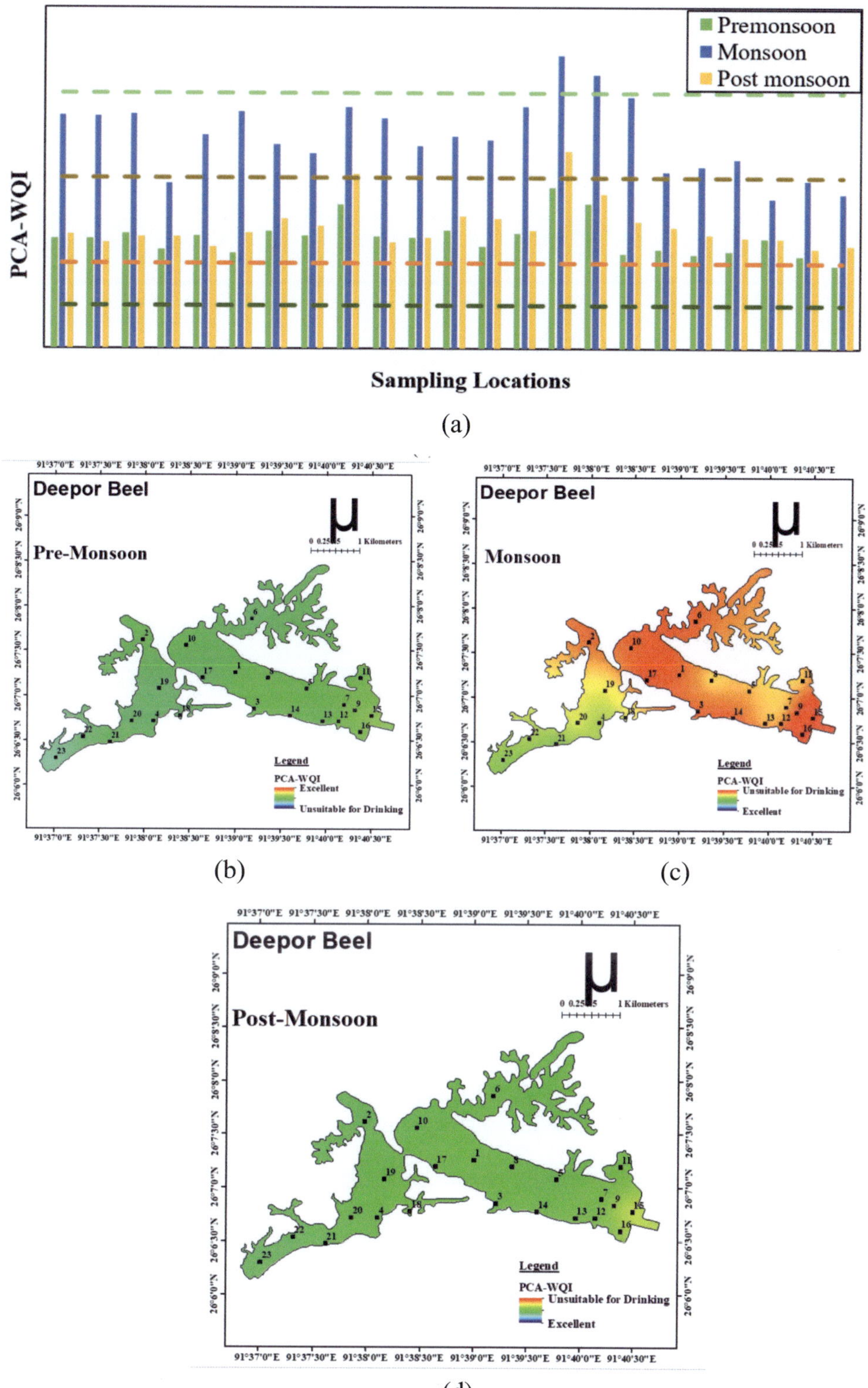

Fig. 4 Spatiotemporal variation of PCA–WQI computed using the proposed methodology

Table 7 Comparative assessment of the percentage distribution of sampling locations for different PCA–WQI classes across three seasons

Classes	Pre-monsoon		Monsoon		Post-monsoon	
	Existing	Proposed	Existing	Proposed	Existing	Proposed
Excellent	0.00	0.00	0.00	0.00	0.00	0.00
Good	46.08	30.44	11.59	10.14	23.91	6.52
Poor	53.92	65.22	27.53	16.67	73.19	88.42
Very poor	0.00	4.34	43.48	43.48	2.18	3.62
Unsuitable for drinking	0.00	0.00	17.40	29.71	0.72	1.44

monsoon period, the water quality by the proposed method was primarily in the 'poor' category and 'unsuitable for drinking' category. Contrarily, only the sites proximate to the landfill displayed 'unsuitable for drinking' water quality. Likewise, the distribution of sites in different categories varied considerably during the post-monsoon period. All variations in the percentage distribution of sampling locations are displayed in Table 7. However, when compared with the spatiotemporal pattern plots of the water quality dataset, it was observed that the results obtained for the proposed PCA–WQI aligned better with the observed values, compared to the PCA–WQI using the current practice. This indicated that the proposed method correlated better with the experimental dataset.

3.3 Sensitivity Analysis

In order to confirm the correctness and reliability of the proposed method, a sensitivity analysis was conducted, the results are depicted in Fig. 5. Figure 5a shows box plots reflecting the statistics of the sensitivity values for each parameter. It was observed that the sensitivity values existed between 0.2% and 2%, showing low sensitivity of input parameters. Only seven parameters (TA, TH, BOD_5, Cl^-, NO_3^-, Na^+ and Ca^{2+}) displayed sensitivity values of more than 1%. The spatiotemporal variability of sensitivity values for all parameters, as shown in Fig. 5b, indicated that sites 1, 5, 6 and 8 showed slight high sensitivity of the seven parameters whose values exceed 1%. Other than these, all sites displayed near-similar sensitivity values for all parameters. This intended that all water quality variables/parameters offered uniform sensitivity for computing PCA–WQI, thus indicating a high level of correctness and reliability for the proposed method.

4 Conclusion

With the United Nations focussing on clean water and sanitation through sustainable development goals (SDG-6), it is vital to monitor the water quality of available freshwater and groundwater sources. In the present chapter, limitations in the current practice of employing multivariate statistics and a proposal for modification in its use were proposed. Critical observations were made, which could be summarised as follows:

- The present method of employing PCA for calculating WQI suffers from limitations of uniformly allocating weights to parameters across all sampling locations irrespective of their characteristics and response to a pollution load. This results in ambiguities, as the sites close to a particular pollution source show different attributes than those close to another pollution source.
- Therefore, the proposed methodology first classifies the sites with similar behaviour and characteristics prior to employing PCA. Hence, it was suggested to first carry out hierarchical clustering of the entire dataset. This would categorise sampling sites into different clusters based on their similarity. The clustering process is followed by employing PCA on individual groups, based on which weights are assigned to the parameters of sites corresponding to that cluster. This process removes ambiguities in allocating equal importance to the parameters for all sampling locations.
- The results of the study indicated that the proposed PCA–WQI proved better than the existing practice. Better correlations were obtained with the experimental/observed water quality in case of the proposed method. The correctness and reliability of the proposed method

Fig. 5 Spatiotemporal variation of PCA–WQI computed using the proposed methodology

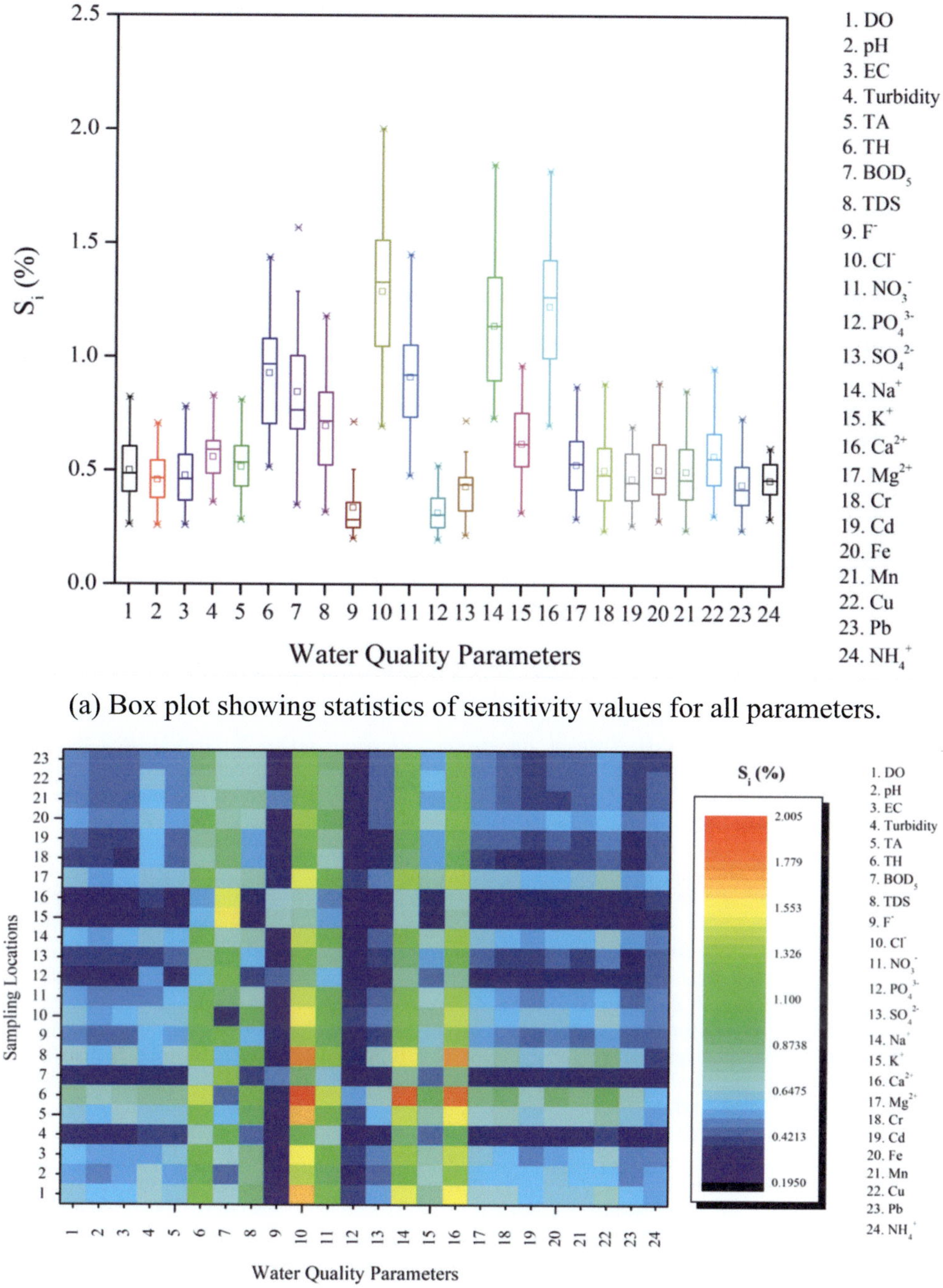

(a) Box plot showing statistics of sensitivity values for all parameters.

(b) Pattern plot showing spatial extent.

were then tested using sensitivity analysis. The results of the sensitivity analysis showed that no parameter had substantially higher sensitivity than others, and all parameters exhibited near-similar low sensitivity values that proved the method's efficacy.

The present study paves the way for employing a novel and scientific technique to assess water quality of a particular water body through a dynamic approach without subjectivity. This further helps in allocating resources by governmental organisations for the restoration process of a water body.

Declaration of Competing Interest The authors declared that they had no known competing interests or personal relationships that could have influenced the work reported in this chapter.

Funding Sources This research did not receive any specific grant from funding agencies in the public, commercial or not-for-profit sources.

Data Availability All data, models and codes generated or used during the study appeared in the text.

Ethical Approval The authors declared that the submitted chapter was an original study. They further asserted that the current research was conducted ethically, and all authors agreed with study's outcome.

References

Adimalla, N., Qian, H., & Li, P. (2020). Entropy water quality index and probabilistic health risk assessment from geochemistry of groundwaters in hard rock terrain of Nanganur county, South India. *Geochemistry, 80*(4), 125544. https://doi.org/10.1016/j.chemer.2019.125544

American Public Health Association (APHA) (2012). *Standard methods for the examination of water and wastewater* (Vol. 2). APHA, American Water Works Association and Water Environment Federation.

Anh, L. D., Loc, H. H., Irvine, K. N., Thanh, T., & Tuong, L. Q. (2021). The waterscape of groundwater exploitation for domestic uses in District 12, Ho Chi Minh City, Vietnam. *Environment, Development and Sustainability, 23*(5), 7652–7669. https://doi.org/10.1007/s10668-020-00938-0

Bansah, K. J., Dumakor-Dupey, N. K., Kansake, B. A., Assan, E., & Bekui, P. (2018). Socioeconomic and environmental assessment of informal artisanal and small-scale mining in Ghana. *Journal of Cleaner Production, 202*, 465–475. https://doi.org/10.1016/j.jclepro.2018.08.150

Bureau of Indian Standards (BIS). (2012). Indian Standard Drinking Water—Specification (Second Revision). IS-10500:2012. BIS.

Castellazzi, P., Longuevergne, L., Martel, R., Rivera, A., Brouard, C., & Chaussard, E. (2018). Quantitative mapping of groundwater depletion at the water management scale using a combined GRACE/InSAR approach. *Remote Sensing of Environment, 205*, 408–418. https://doi.org/10.1016/j.rse.2017.11.025

Chen, M., & Chen, H. (2021). Spatiotemporal coupling measurement of industrial wastewater discharge and industrial economy in China. *Environmental Science and Pollution Research, 28*(34), 46319–46333. https://doi.org/10.1007/s11356-021-14743-3

Collins, A. L., Newell Price, J. P., Zhang, Y., Gooday, R., Naden, P. S., & Skirvin, D. (2018). Assessing the potential impacts of a revised set of on-farm nutrient and sediment 'basic' control measures for reducing agricultural diffuse pollution across England. *Science of the Total Environment, 621*, 1499–1511. https://doi.org/10.1016/j.scitotenv.2017.10.078

Dalal, S. G., Shirodkar, P. V., Jagtap, T. G., Naik, B. G., & Rao, G. S. (2010). Evaluation of significant sources influencing the variation of water quality of Kandla creek, Gulf of Katchchh, using PCA. *Environmental Monitoring and Assessment, 163*(1–4), 49–56. https://doi.org/10.1007/s10661-009-0815-y

Dalin, C., Wada, Y., Kastner, T., & Puma, M. J. (2017). Groundwater depletion embedded in international food trade. *Nature, 543*(7647), 700–704. https://doi.org/10.1038/nature21403

Dash, S., Borah, S. S., & Kalamdhad, A. (2018). Monitoring and assessment of Deepor Beel water quality using multivariate statistical tools. *Water Practice and Technology, 13*(4), 893–908. https://doi.org/10.2166/wpt.2018.098

Dash, S., Borah, S. S., & Kalamdhad, A. (2019). A modified indexing approach for assessment of heavy metal contamination in Deepor Beel, India. *Ecological Indicators, 106*, 105444. https://doi.org/10.1016/j.ecolind.2019.105444

Dash, S., Borah, S. S., & Kalamdhad, A. S. (2020a). Application of environmetrics tools for geochemistry, water quality assessment and apportionment of pollution sources in Deepor Beel, Assam, India. *Water Practice and Technology, 15*(4), 973–992. https://doi.org/10.2166/wpt.2020.078

Dash, S., Borah, S. S., & Kalamdhad, A. S. (2020b). Study of the limnology of wetlands through a one-dimensional model for assessing the eutrophication levels induced by various pollution sources. *Ecological Modelling, 416*, Article 108907. https://doi.org/10.1016/j.ecolmodel.2019.108907

Dash, S., & Kalamdhad, A. S. (2021a). Hydrochemical dynamics of water quality for irrigation use and introducing a new water quality index incorporating multivariate statistics. *Environmental Earth Sciences, 80*(3), 73. https://doi.org/10.1007/s12665-020-09360-1

Dash, S., & Kalamdhad, A. S. (2021b). Science mapping approach to critical reviewing of published literature on water quality indexing. *Ecological Indicators, 128*, Article 107862. https://doi.org/10.1016/j.ecolind.2021.107862

Dash, S., & Kalamdhad, A. S. (2021c). Discussion on the existing methodology of entropy-weights in water quality indexing and proposal for a modification of the expected conflicts. *Environmental Science and Pollution Research, 28*(38), 53983–54001. https://doi.org/10.1007/s11356-021-14482-5

Gao, Y., Qian, H., Ren, W., Wang, H., Liu, F., & Yang, F. (2020). Hydrogeochemical characterization and quality assessment of groundwater based on integrated-weight water quality index in a concentrated urban area. *Journal of Cleaner Production, 260*, Article 121006. https://doi.org/10.1016/j.jclepro.2020.121006

Howladar, M. F., Al Numanbakth, M. A., & Faruque, M. O. (2018). An application of water quality index (WQI) and multivariate statistics to evaluate the water quality around Maddhapara granite mining industrial area, Dinajpur, Bangladesh. *Environmental Systems Research, 6*(1), 13. https://doi.org/10.1186/s40068-017-0090-9

Islam, A. R. M. T., Ahmed, N., Bodrud-Doza, M., & Chu, R. (2017). Characterizing groundwater quality ranks for drinking purposes in Sylhet district, Bangladesh, using entropy method, spatial autocorrelation index, and geostatistics. *Environmental Science and Pollution Research, 24*(34), 26350–26374. https://doi.org/10.1007/s11356-017-0254-1

Jain, M., Fishman, R., Mondal, P., Galford, G. L., Bhattarai, N., & Naeem, S., et al. (2021). Groundwater depletion will reduce cropping intensity in India. *Science Advances, 7*(9), PMC7904249. https://doi.org/10.1126/sciadv.abd2849

Kaiser, H. F. (1960). The application of electronic computers to factor analysis. *Educational and Psychological Measurement, 20*(1), 141–151. https://doi.org/10.1177/001316446002000116

Kumar, D., & Alappat, B. J. (2009). NSF-water quality index: Does it represent the experts' opinion? *Practice Periodical of Hazardous, Toxic, and Radioactive Waste Management, 13*(1), 75–79. https://doi.org/10.1061/(ASCE)1090-025X(2009)13:1(75)

Lermontov, A., Yokoyama, L., Lermontov, M., & Machado, M. A. S. (2009). River quality analysis using fuzzy water quality index: Ribeira do Iguape river watershed, Brazil. *Ecological Indicators, 9*(6), 1188–1197. https://doi.org/10.1016/j.ecolind.2009.02.006

Lund Schlamovitz, J., & Becker, P. (2021). Differentiated vulnerabilities and capacities for adaptation to water shortage in Gaborone, Botswana. *International Journal of Water Resources Development, 37*(2), 278–299. https://doi.org/10.1080/07900627.2020.1756752

Mostafaei, A. (2014). Application of multivariate statistical methods and water-quality index to evaluation of water quality in the Kashkan

River. *Environmental Management, 53*(4), 865–881. https://doi.org/10.1007/s00267-014-0238-6

Nasiri, F., Maqsood, I., Huang, G., & Fuller, N. (2007). Water quality index: A fuzzy river-pollution decision support expert system. *Journal of Water Resources Planning and Management, 133*(2), 95–105. https://doi.org/10.1061/(ASCE)0733-9496(2007)133:2(95)

Ngasala, T. M., Gasteyer, S. P., Masten, S. J., & Phanikumar, M. S. (2019). Linking cross contamination of domestic water with storage practices at the point of use in urban areas of Dar es Salaam, Tanzania. *Journal of Environmental Engineering, 145*(5), 04019017. https://doi.org/10.1061/(ASCE)EE.1943-7870.0001516

Patil, V. B. B., Pinto, S. M., Govindaraju, T., Hebbalu, V. S., Bhat, V., & Kannanur, L. N. (2020). Multivariate statistics and water quality index (WQI) approach for geochemical assessment of groundwater quality—A case study of Kanavi Halla sub-basin, Belagavi, India. *Environmental Geochemistry and Health, 42*(9), 2667–2684. https://doi.org/10.1007/s10653-019-00500-6

Preisner, M. (2020). Surface water pollution by untreated municipal wastewater discharge due to a sewer failure. *Environmental Processes, 7*(3), 767–780. https://doi.org/10.1007/s40710-020-00452-5

Proulx, F., Rodriguez, M. J., Sérodes, J. B., & Miranda, L. F. (2010). Factors influencing public perception and use of municipal drinking water. *Water Supply, 10*(3), 472–485. https://doi.org/10.2166/ws.2010.511

Ray, S., Mishra, A. K., & Kalamdhad, A. S. (2021a). Influence of real and synthetic municipal solid waste leachates on consolidation and shear strength behaviour of bentonites. *Environmental Science and Pollution Research, 28*(24), 30975–30985. https://doi.org/10.1007/s11356-021-12863-4

Ray, S., Mishra, A. K., Kalamdhad, A. S., & Reddy, C. V. (2021b). Impact of real and simulated municipal solid waste leachates on the hydraulic and swelling behaviour of bentonites for landfill application. *Environmental Monitoring and Assessment, 193*(11), 701. https://doi.org/10.1007/s10661-021-09510-3

Salehi, M. (2022). Global water shortage and potable water safety; today's concern and tomorrow's crisis. *Environment International, 158*, Article 106936. https://doi.org/10.1016/j.envint.2021.106936

Singaraju, S., Pasupuleti, S., Hernandez, E. A., & Uddameri, V. (2018). Prioritizing groundwater monitoring in data sparse regions using Atanassov intuitionistic fuzzy sets (A-IFS). *Water Resources Management, 32*(4), 1483–1499. https://doi.org/10.1007/s11269-017-1883-3

Singh, S. K., Singh, P., & Gautam, S. K. (2016). Appraisal of urban lake water quality through numerical index, multivariate statistics and earth observation data sets. *International Journal of Environmental Science and Technology, 13*(2), 445–456. https://doi.org/10.1007/s13762-015-0850-x

Sutadian, A. D., Muttil, N., Yilmaz, A. G., & Perera, B. J. C. (2017). Using the analytic hierarchy process to identify parameter weights for developing a water quality index. *Ecological Indicators, 75*, 220–233. https://doi.org/10.1016/j.ecolind.2016.12.043

Tabachnick, B. G., Fidell, L. S., & Ullman, J. B. (2007). *Using multivariate statistics* (Vol. 5). Pearson.

Varol, S., & Davraz, A. (2015). Evaluation of the groundwater quality with WQI (water quality index) and multivariate analysis: A case study of the Tefenni plain (Burdur/Turkey). *Environmental Earth Sciences, 73*(4), 1725–1744. https://doi.org/10.1007/s12665-014-3531-z

Wanda, E. M. M., Gulula, L. C., & Kushe, J. (2014). An assessment of effectiveness of the Lunyangwa River catchment co-management model in Mzuzu City, Northern Malawi. *Physics and Chemistry of the Earth, Parts a/b/c, 72–75*, 96–103. https://doi.org/10.1016/j.pce.2014.10.003

Wang, D., Singh, V. P., Zhu, Y., & Wu, J. (2009). Stochastic observation error and uncertainty in water quality evaluation. *Advances in Water Resources, 32*(10), 1526–1534. https://doi.org/10.1016/j.advwatres.2009.07.004

Wilks, D. (2011). *Statistical methods in the atmospheric sciences* (3rd ed.). Academic Press.

Yaseen, Z. M., Ramal, M. M., Diop, L., Jaafar, O., Demir, V., & Kisi, O. (2018). Hybrid adaptive neuro-fuzzy models for water quality index estimation. *Water Resources Management, 32*(7), 2227–2245. https://doi.org/10.1007/s11269-018-1915-7

Simultaneous Determination of Heavy Metal Ions in Water Samples Using Anodic Stripping Voltammetry

Monika Antil and Babankumar S. Bansod

1 Introduction

Environmental pollution, a global challenge, has been exacerbated by the pervasive presence of heavy metal ions. These elements, characterised by their non-biodegradability and potent toxicity, pose a significant threat to ecological balance and human health (Hu et al., 2022). Lead ($Pb2^+$) and cadmium ($Cd2^+$) ions, primary pollutants released by industrial and agricultural activities (Wang & Yang, 2023), exert significant deleterious effects on human health, notably impacting hematopoietic, cardiovascular, nervous and immune systems (Bolisetty & Mezzenga, 2016). Chronic lead exposure disrupts neurometabolic function in mice, impairing energy production and cognitive processes. A team of senior scientists at the Centre for Cellular and Molecular Biology (CCMB), Hyderabad, India, demonstrated that chronic lead exposure adversely affects the brain's ability to metabolise nutrients into energy, consequently impacting cognitive function (Sai Gopal, 2023).

Precise and sensitive detection of cadmium and lead ions is paramount for safeguarding environmental and human health. Spectroscopic techniques, such as atomic absorption spectroscopy (AAS) (Janusa & Beck, 2002), atomic emission spectrometry (AES) (Zhang et al., 2021), inductively coupled plasma mass spectrometry (ICP-MS) (Skrzydlewska et al., 2003) and X-ray fluorescence spectrometry (XRF) (Radu & Diamond, 2009), have proven to be effective tools for quantifying these toxic metals. Conventional ion detection methods, while precise, often necessitate intricate sample pre-treatment, specialised and costly instrumentation and highly skilled personnel. These requirements contribute to low efficiency, elevated costs and limitations in on-site deployment. Electrochemical detection emerges as a viable alternative, offering several advantages. These include simplified instrumentation, user-friendly operation, high sensitivity, selective ion recognition, portability and rapid analysis capabilities (Baghayeri et al., 2018; Chen, 2021; Li et al., 2016; Wang et al., 2022).

Recent research efforts have focused on enhancing the limits of detection (LOD) and sensitivity of heavy metal ion analysis. Despite the ability to identify the presence of heavy metals, the accurate determination of specific target metal ions remains a significant challenge because of the complex interactions between coexisting metal species. Additionally, the development of electrode materials that exhibit both long-term stability and high sensitivity, while simultaneously achieving selectivity in real-world sample matrices, continues to be an obstacle. The presence of organic interferents, such as humic acid, surfactants and colloids, further complicates the analysis by competing for active sites, altering stripping potentials and fouling the electrode surface. A comprehensive understanding of the electrochemical sensing process, including its reaction kinetics, necessitates additional analytical and experimental data.

Biyani and Radhika (2017) developed a disposable electrode-printed chip for the simultaneous detection of lead and cadmium ions in environmental samples. The system demonstrated LOD of four *parts-per-billion* (ppb) and 2.6 ppb for lead and cadmium, respectively. However, several limitations were identified such as sample preparation, interference from the presence of other heavy metal ions that interfere with the analysis by competing for active sites on electrode surface, and the supporting electrolyte

M. Antil · B. S. Bansod (✉)
Department of Materials Science and Sensor Applications MSSA, CSIR-Central Scientific Instruments Organisation, Chandigarh, India
e-mail: scientist_babankumar@csio.res.in

Academy of Scientific and Innovative Research (AcSIR), Haziabad, India

© Capital Publishing Company, New Delhi, India 2025

I. Haq et al. (eds.), *Water and Wastewater*, https://doi.org/10.1007/978-3-031-94253-2_6

contributing an additional peak, thus complicating data analysis. Overall, the method suffers from inadequate sensitivity and selectivity.

Pandey et al. (2019) developed an electrochemical sensor for the detection of cadmium and lead ions. The sensor employed a glassy carbon electrode (GCE) modified with graphene oxide (GO) dispersed in the ionic liquid 1-butyl-3-methylimidazolium hexafluorophosphate ($BMIM\text{-}PF_6$) and stabilised with a Nafion binder. This sensor exhibited high sensitivity, enabling the detection of cadmium and lead at concentrations as low as 0.3 ppb and 0.42 ppb, respectively. Interference studies revealed that the sensor's performance was influenced by the presence of other ions. Notably, copper ($Cu2^+$) ions were found to significantly interfere with the detection of both cadmium and lead, leading to increased peak currents and reduced sensitivity (Pandey, 2019).

Baghayeri et al. (2019) employed a GCE coated with poly (amidoamine) functionalised magnetic graphene oxide (GO-Fe_3O_4-PAMAM) for the detection of cadmium and lead ions, achieving LOD of 0.070 ppb and 0.130 ppb, respectively. However, the presence of $Cu2^+$ ions was found to significantly suppress the peak currents of $Cd2^+$ (17.6% reduction) and Pb^{2+} (13.4% reduction), compromising sensitivity (Baghayeri et al., 2019).

To address these limitations, this study focusses on a simplified approach utilising an unmodified standard gold electrode. This strategy is anticipated to enhance ruggedness, stability and portability while reducing reagent consumption and improving overall sensitivity for cadmium and lead ion detection.

2 Material and Instrumentation Used

2.1 Instrumentation

Electrochemical measurements, including square wave anodic stripping voltammetry (SWASV), were performed in a conventional three-electrode glass cell equipped with an airtight Teflon cap. An Autolab potentiostat (PGSTAT302N; CH Instruments, the Netherlands) was employed for data acquisition.

2.2 Materials (Chemical/Reagent)

All high-purity reagents were procured of make Sigma-Aldrich from Oswal Scientific Limited Chandigarh, unless otherwise specified and employed without additional purification. Deionised water (resistivity $\geq$ 18.2 mega ohms per centimetre [MΩ/cm]) was used for the preparation of all samples and stock solutions. Standard solutions of lead and cadmium were purchased from Inorganic Ventures Inc. at 300 Technology Dr, Christiansburg, VA 24,073, USA. Nitric acid (69% purity) and ethanol (95% purity) were obtained from a local Sigma-Aldrich distributor in India.

3 Results and Discussion

The anodic stripping voltammetry (ASV) was employed to determine the cadmium and lead ions' sensitively. Accordingly, the stripping voltammetry parameters were tuned.

3.1 Optimisation of the Pre-concentration of Lead and Cadmium Ions

This research investigates the optimisation of pre-concentration parameters for lead and cadmium ions. Pre-concentration is critical in enhancing the sensitivity and LOD of analytical techniques employed for analysis of heavy metal.

The present study systematically evaluates the influence of pH, electrolyte concentration, deposition time and electrode material on the pre-concentration efficiency of lead and cadmium ions. Anodic stripping voltammetry was utilised as an electrochemical technique to assess the impact of these parameters. Experiments were conducted using synthetic standard solutions containing known concentrations of lead and cadmium to establish optimal pre-concentration conditions.

The detailed findings of this investigation are presented in the following section.

3.2 Influence of Deposition Potential on Stripping Peak Characteristics and Analytical Performance of Lead and Cadmium

A comprehensive investigation was conducted to assess the impact of varying deposition potentials on the stripping voltammetric determination of lead and cadmium. These metal ions were selected as model analytes because of their environmental and analytical significance. Solutions containing known concentrations of lead and cadmium were subjected to a range of deposition potentials during the accumulation phase. The deposition process involved the reduction of metal ions onto electrode surface. Subsequently, anodic stripping voltammetry was employed to oxidise the accumulated metal species, generating corresponding current peaks. The influence of deposition potential on peak current intensity was evaluated, with optimal conditions determined at -1.0 V for lead and -1.2 V for cadmium. These findings provide a foundation for subsequent analytical applications.

3.3 Influence of Deposition Time on Stripping Peak Characteristics and Analytical Performance of Lead and Cadmium

A systematic investigation of deposition time was conducted within the range of 60–350 s, incrementing by 60-s intervals, to ascertain its influence on the stripping process. The results indicated a direct correlation between deposition time and the intensity of stripping peaks for both lead and cadmium ions, as reported previously (Shen & Zhang, 2017). Optimal deposition period was identified to maximise signal response. However, excessive deposition period led to peak broadening and compromised analytical precision. Consequently, the deposition time was optimised to achieve a balance between sensitivity and peak shape.

4 Linearity and Calibration Response

This study investigates the linearity of sensor response to analyse concentration for lead and cadmium. To establish this relationship, a series of standard solutions with varying concentrations of target analytes was prepared. The resulting sensor responses were plotted against the corresponding analyte concentrations (cadmium and lead) to generate a calibration curve (Figs. 1 and 2). The linearity of this plot indicates a directly proportional relationship between analyte concentration and sensor signal. The slope and intercept of calibration curve are crucial parameters for quantitative analysis of unknown samples (Hassan & Gaber, 2020). The slope of a calibration plot quantitatively represents the analytical sensitivity of a method, while the intercept provides information about the method's baseline offset or background signal.

4.1 Simultaneous Response for Concentrations of Cadmium and Lead Ions

Simultaneous response defined the ability of an analytical method to accurately determine the concentrations of cadmium and lead ions in a sample without prior separation. In the present study, this is achieved by exploiting the unique analytical signals generated by each ion (Fig. 3).

The linear expression of lead and cadmium is expressed in Equation A1:

$$\text{A1.}Cd^{2+}: I = 31.89x + 5.1,$$

$$Pb^{2+}: I = 59.09 + 4.56.$$

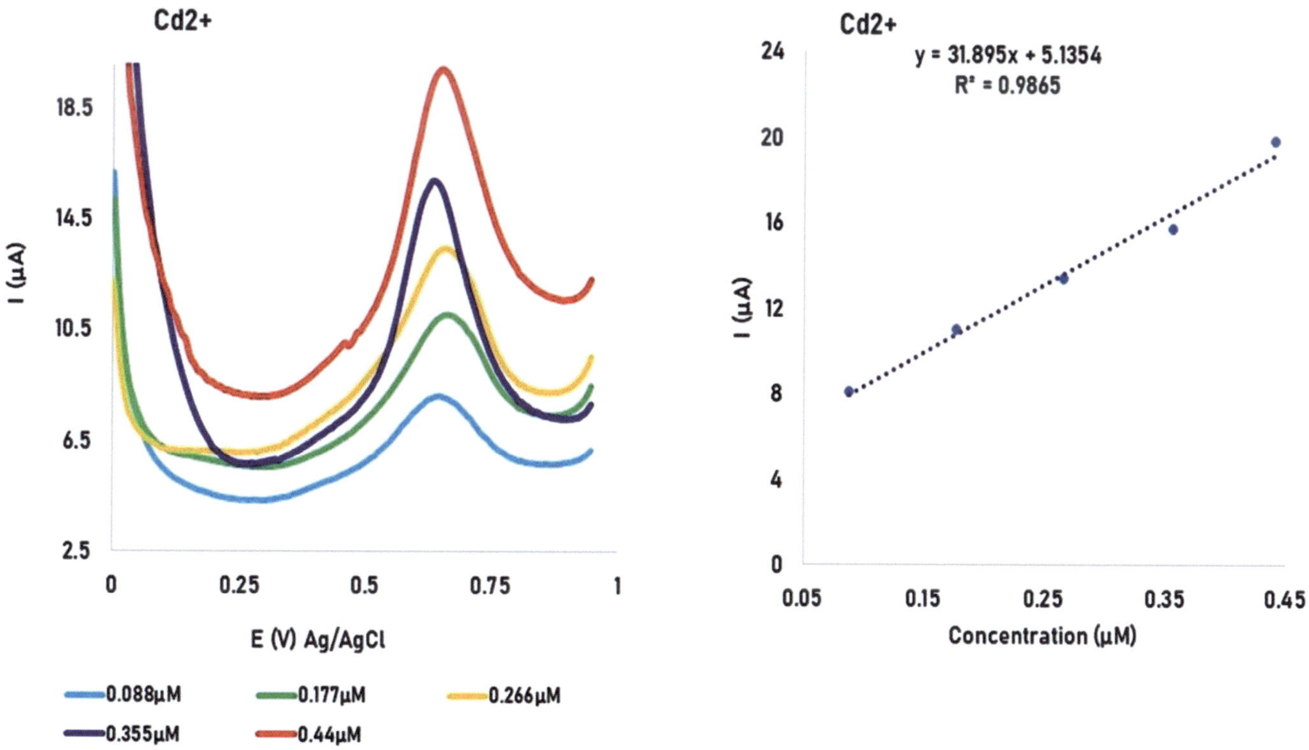

Fig. 1 Response of anodic stripping voltammetry for a concentration range of 0.088–0.44 µM (10–50 ppb) for cadmium (Cd^{2+}) ions

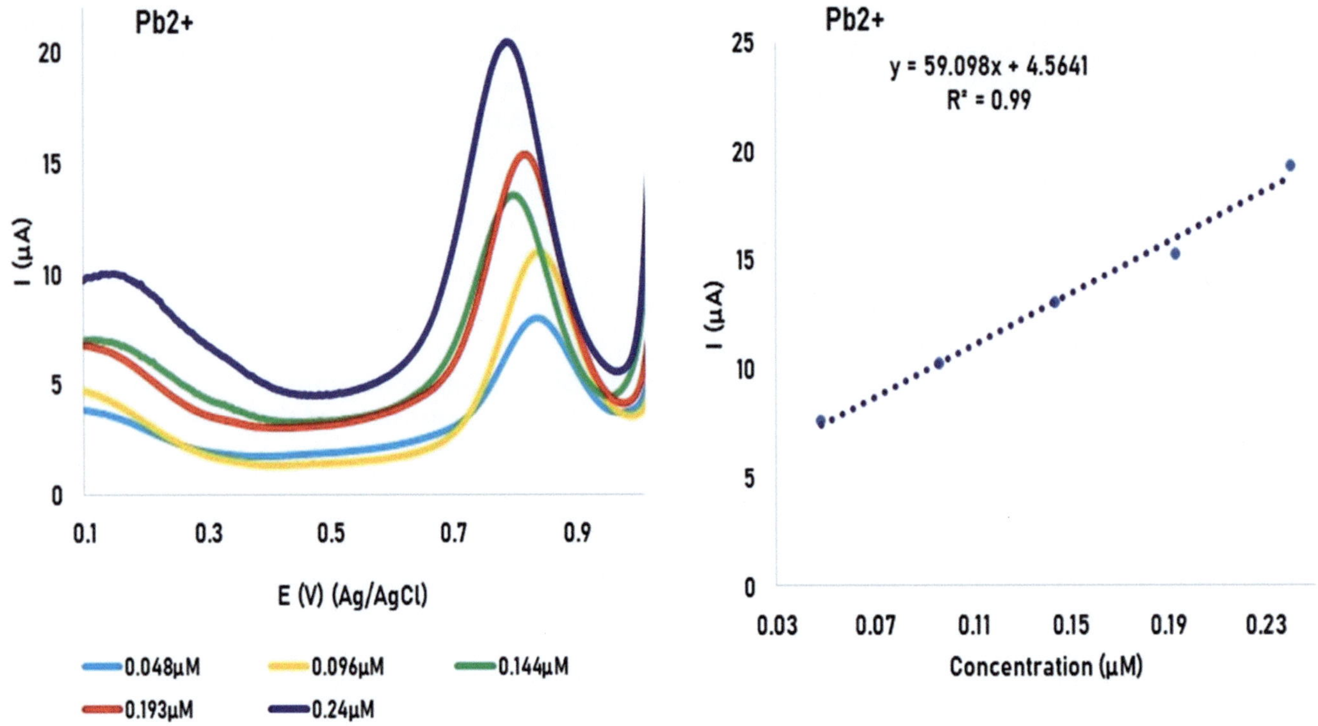

Fig. 2 Response of anodic stripping voltammetry for a concentration range of 0.048–0.24 μM (10–50 ppb) for lead (Pb^{2+}) ions

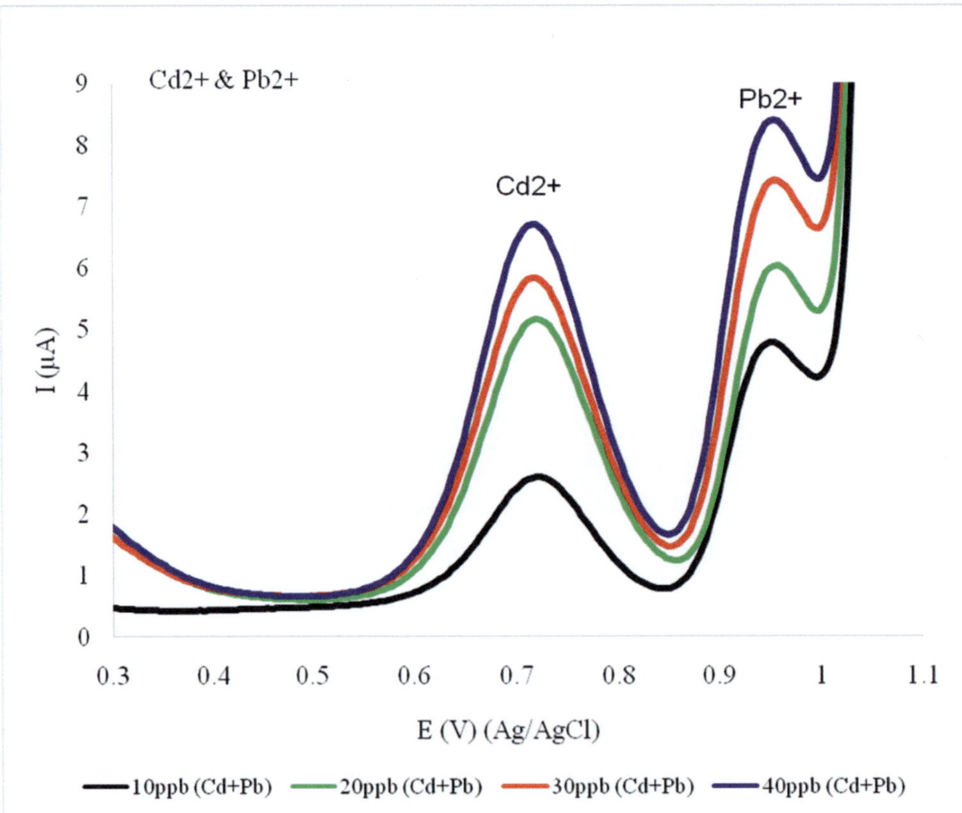

Fig. 3 Response of anodic stripping voltammetry for a concentration range of cadmium (Cd^{2+}) and lead (Pb^{2+}) ions (10–40 ppb)

As depicted in Figs. 1 and 2, the calibration curves for cadmium and lead ions exhibit excellent linearity. The corresponding sensitivity values (slopes) and regression coefficients (*R*2) demonstrate robust and accurate quantification capabilities for both analytes.

Linear regression analysis yielded the following equations:

- Cadmium: I (μA) = 31.895 (μA/μM) × [Cd] + 5.14, $R2 = 0.986$,
- Lead: I (μA) = 59.098 (μA/μM) × [Pb] + 4.78, $R2 = 0.99$.

LOD, signifying the lowest analyte concentration measurable with confidence, was determined using Eq. A2:

$$\text{A2 Lo D} = 3.3 \times \text{standard deviation/slope}.$$

The method demonstrated LOD for lead as 7×10^{-3} μM and that for cadmium as 5×10^{-4} μM (Table 1). These values meet the global criteria for drinking water established by the World Health Organization (WHO) and the US Environmental Protection Agency (US EPA) (Paul, 2017; Shrivastava, 2011).

Notably, the achieved LOD values are significantly lower than those reported in previous studies (Table 2), suggesting the reliability of the present method for heavy metal analysis in real water samples.

5 Conclusion

A gold microelectrode was employed as an electrochemical sensor for simultaneous, sensitive and robust determination of lead and cadmium ions using anodic stripping voltammetry. The sensor exhibited excellent selectivity for target analytes, enabling their simultaneous quantification in aqueous solutions down to LOD of 7×10^{-3} μM for lead and 5×10^{-4} μM for cadmium. The utilisation of microelectrodes significantly enhanced the system's sensitivity while minimising sample volume requirements.

Table 1 Limit of detection for lead (Pb^{2+}) and cadmium (Cd^{2+}) ions

Sr. no	Detecting species	Linear range	Regression coefficient	LOD
1	Pb^{2+}	0.048–0.24 μM	0.99	7×10^{-3} μM
2	Cd^{2+}	0.088–0.44 μM	0.986	5×10^{-4} μM

Note LOD: limit of detection

Table 2 Comparison of reported work for cadmium (Cd^{2+}) and lead (Pb^{2+}) sensing using electrochemical techniques

Technique	Analytes	Sensing material	Range	LOD	References
Differential pulse voltammetry	Cd(II) Pb(II)	Yb-BTC/glassy carbon electrode	0–50 ppb	3 ppb 1.6 ppb	Manh et al., (2021)
Square wave anodic stripping voltammetry	Cd(II) Pb(II)	Glutathione-coated Fe_3O_4-modified electrode	0.5–100 ppb	0.172 ppb 0.182 ppb	Baghayeri et al. (2018)
Square wave anodic stripping voltammetry	Cd(II) Pb(II)	Fe_3O_4/F-MWCNTs/glassy carbon electrode	56.2–3,372 ppb 103.6–6,216 ppb	5.62 ppb 16.58 ppb	Wu et al. (2019)
Square wave anodic stripping voltammetry	Cd(II) Pb(II)	Nanoplated Fe_3O_4-modified electrode	11.24–224.8 ppb 8.3–870 ppb	17.53 ppb 3.15 ppb	Li and Yao (2015)
Differential pulse voltammetry	Cd(II) Pb(II)	GA-UiO-66-NH_2/glassy carbon electrode	6.745–337.2 ppb 2.072–0.828 ppb	2.248 ppb 0.3108 ppb	Lu (2019)
Square wave anodic stripping voltammetry	Cd(II) Pb(II)	g-C_3N_4/CNT/ NH_2-MIL-88(Fe)-modified electrode	13.49 ppb–0.6745 ppm 4.144 ppb–1.2432 ppm	4.45 ppb 1.57 ppb	Chen (2021)
Differential pulse voltammetry	Cd(II) Pb(II)	Tapioca-derived carbon dots-modified electrode	1.01–0.27 ppm 0.01–0.27 ppm	2.8 ppb 4.2 ppb	Pudza (2020)
Square wave anodic stripping voltammetry	Cd(II) Pb(II)	Gold (Au)-coated nanoparticles-modified electrode	10–50 ppb	0.067 ppb 1.49 ppb	Present Work

Note LOD: limit of detection

This anodic stripping voltammetry-based platform holds immense potential for real-time, on-site monitoring of heavy metal contamination in water bodies. Its ability to simultaneously analyse multiple analytes offers a cost-effective and time-efficient approach to water quality assessment. Furthermore, the compact microelectrode-based set-up facilitates reduced reagent consumption and enhanced portability, making it suitable for field applications and remote monitoring.

Acknowledgements The authors are grateful to the Academy of Scientific and Innovative Research (AcSIR), Ghaziabad, India, and the Director CSIR-CSIO Chandigarh, India, for facilitating the reported research work.

References

Baghayeri, M., Amiri, A., Maleki, B., Alizadeh, Z., & Reiser, O. (2018). A simple approach for simultaneous detection of cadmium(II) and lead(II) based on glutathione coated magnetic nanoparticles as a highly selective electrochemical probe. *Sensors Actuators b: Chemical, 273*, 1442–1450. https://doi.org/10.1016/j.snb.2018.07.063

Baghayeri, M., Alinezhad, H., Fayazi, M., Tarahomi, M., Ghanei-Motlagh, R., & Maleki, B. (2019). A novel electrochemical sensor based on a glassy carbon electrode modified with dendrimer functionalized magnetic graphene oxide for simultaneous determination of trace Pb(II) and Cd(II). *Electrochimica Acta*, 80–88. https://doi.org/10.1016/j.electacta.2019.04.180

Biyani, M., & Radhika, B. (2017). DEP-on-go for simultaneous sensing of multiple heavy metals pollutants in environmental samples. *Sensors*. https://doi.org/10.3390/s17010045

Bolisetty, S., & Mezzenga, R. (2016). For universal water purification. *Nature Nanotechnology*, 365−371. https://doi.org/10.1038/nnano.2015.310

Chen, S. (2021). Simultaneous electrochemical sensing of heavy metal ions based on g-C_3N_4/CNT/NH_2-MIL-88(Fe) nanocomposite. *Analytical Methods*. https://doi.org/10.1039/D1AY01682B

Hassan, K. M., & Gaber, S. E. (2020). Single and simultaneous voltammetric sensing of lead(II), cadmium(II) and zinc(II) using a bimetallic Hg-Bi supported on poly(1,2-diaminoanthraquinone)/glassy carbon modified electrode. *Sensing and Bio-Sensing Research, 29*, Article 100369. https://doi.org/10.1016/j.sbsr.2020.100369

Hu, H., Xie, B., Lu, Y., & Zhu, J. (2022). Advances in electrochemical detection electrodes for As(III). *Nanomaterials*, *12*, 781. https://doi.org/10.3390/nano12050781

Janusa, M. A., & Beck, J. N. (2002). Recent applications of flame atomic absorption spectrometry to environmental measurements. *Applied Spectroscopy Reviews, 37*(2), 137–186. https://doi.org/10.1081/ASR-120006043

Li, W. J., & Yao, X.-Z. (2015). Fe_3O_4 with novel nanoplate-stacked structure: Surfactant-free hydrothermal synthesis and application in detection of heavy metal ions. *Journal of Electroanalytical Chemistry, 749*, 75–82. https://doi.org/10.1016/j.jelechem.2015.04.038

Li, N., Zhang, D., Zhang, Q., Lu, Y., Jiang, J., Liu, G. L., & Liu, Q. (2016). Combining localized surface plasmon resonance with anodic stripping voltammetry for heavy metal ion detection. *Sensor and Actuators B Chemical, 231*, 349–356. https://doi.org/10.1016/j.snb.2016.03.042

Lu, M. (2019). Graphene aerogel-metal-organic framework-based electrochemical method for simultaneous detection of multiple heavy-metal ions. *Analytical Chemistry, 91*, 888–895. https://doi.org/10.1021/acs.analchem.8b03764

Manh, B., Dau, N., Ngoc, T., & Nga, V. T. (2021). Novel nanoscale Yb-MOF used as a highly efficient electrode for simultaneous detection of heavy metal ions. *Journal of Material Science, 56*, 8172–8185. https://doi.org/10.1007/s10853-021-05815-3

Pandey Kumar, S., Sachan Sadhana, S., & Sunil Kumar, S. (2019). Ultratrace sensing of cadmium and lead by square wave anodic stripping voltammetry using ionic liquid modified graphene oxide. *Materials Science for Energy Technologies*, 667–675. https://doi.org/10.1016/j.mset.2019.09.004

Paul, D. (2017). Research on heavy metal pollution of river Ganga: A review. *Annals of Agrarian Science, 15*(2), 278–286. https://doi.org/10.1016/j.aasci.2017.04.001

Pudza, M. Y. (2020). Selective and simultaneous detection of cadmium, lead and copper by tapioca-derived carbon dot–modified electrode. *Environmental Science and Pollution Research, 27*(12): 13315–13324. https://doi.org/10.1007/s11356-020-07695-7

Radu, T., & Diamond, D. (2009). Comparison of soil pollution concentrations determined using AAS and portable XRF techniques. *Journal of Hazardous Material, 171*, 1168–1171. https://doi.org/10.1016/j.jhazmat.2009.06.062

Sai Gopal, M. (2023). Lead toxicity major health concern. *Telangana Today* (E-paper).

Shen, L.-L., & Zhang, G.-R. (2017). Modifier-free microfluidic electrochemical sensor for heavy-metal. *ACS Omega*, 4593−4603. https://doi.org/10.1021/acsomega.7b00611

Shrivastava, A. (2011). Methods for the determination of limit of detection and limit of quantitation of the analytical methods. *Chronicles of Young Scientists, 2*(1), 21–25. https://doi.org/10.4103/2229-5186.79345

Skrzydlewska, E., Balcerzak, M., & Vanhaecke, F. (2003). Determination of chromium, cadmium and lead in food-packaging materials by axial inductively coupled plasma time-of-flight mass spectrometry. *Analytical Chemical Acta*, 191−202. https://doi.org/10.1016/S0003-2670(02)01527-1

Wang, Y., Wu, X., Sun, J., Wang, C., Zhu, G., Bai, L., Jiang, Z., & Zhang, W. (2022). Stripping voltammetric determination of cadmium and lead ions based on a bismuth oxide surface-decorated nanoporous bismuth electrode. *Electrochemistry Communications, 136*, Article 107233. https://doi.org/10.1016/j.elecom.2022.107233

Wang, M., & Yang, L. (2023). The evaluation and sources of heavy metal anomalies in the surface soil of Eastern Tibet. *Minerals, 13*(1), 86. https://doi.org/10.3390/min13010086

Wu, W., Jia, M., Wang, Z., Zhang, W., Zhang, Q., Liu, G., Zhang, Z., & Li, P. (2019). Simultaneous voltammetric determination of cadmium(II), lead(II), mercury(II), zinc(II), and copper(II) using a glassy carbon electrode modified with magnetite (Fe_3O_4) nanoparticles and fluorinated multiwalledcarbon nanotubes. *Microchimica Acta, 186*(97), 1–10. https://doi.org/10.1007/s00604-018-3216-5

Zhang, F., Wei, N., Wang, J., Liu, Q., Zhaxi, C., Li, J., & Qiu, C. (2021). Determination of available lead in greenhouse soil by solid phase dispersion extraction coupled with atomic absorption spectrometry based on primary secondary amine. *IOP Conference Series. Earth and Environmental Science, 769*, 022056. https://doi.org/10.1088/1755-1315/769/2/022056

Decontamination of Carbon-Based Nanomaterials from Contaminated Water Samples by Employing Microemulsion and Cloud Point Extraction Techniques

Lavi Dhiman, Dinesh Bhardwaj, Vikas Kumar Gahlot, Shalini Anand, Santosh Kumar Singh, and Pramod Kumar Rai

1 Introduction

Nowadays, nanomaterials (NMs) are widely used in every sector of industry. Nanomaterials have a size in the range of 1–100 nm. Owing to their size and possession of unique properties, they have a wide range of applications across various industries, viz. electronics (Gong & Cheng, 2017), biomedical applications (Xiong et al., 2018), catalysis (Gómez-López et al., 2020), energy storage (Azim et al., 2022), textiles and fabrics (Shah et al., 2022), automotive (Imran et al., 2022), cosmetics (Gupta et al., 2022) and construction materials (Shilar et al., 2022). In the wake of their extensive utilisation, nanomaterials are produced in large quantities, and after their use, they are discharged into the environment through avenues such as atmospheric emissions, aquatic systems and wastewater outlets. Owing to such dissemination, these nanomaterials have the potential to induce toxic effects on various organisms inhabiting these ecosystems (Giese et al., 2018) as depicted in Fig. 1.

A study recently stated that metal oxides, viz. silicon oxide (SiO_2), iron oxide (Fe_3O_4) and zinc oxide (ZnO) *nanoparticles*, in combination interrupted photosynthetic action, persuaded oxidative stress in cell system and triggered antioxidant defence systems in algae *Scenedesmus obliquus* causing the entire cell toxicity (Das et al., 2023a). Nanomaterials released into the environment from industrial effluent, sewage waste and many more sources elicit concerns about their potential toxic effects on living organisms, including humans. Owing to their diminutive dimensions and distinctive characteristics, nanomaterials possess the capacity to engage with biological systems in a manner unattainable by conventional bulk materials. Recent studies have indicated that nanomaterials can cause major lung damage. Inhalation of certain nanomaterials, especially those with a high aspect ratio, such as nanotubes or nanofibres, can lead to inflammation and lung damage (Zhou et al., 2023). The presence of nanomaterials carries the potential to elevate the risk of cancer and other genetic disorders because of their ability to interact with cellular components. Certain nanomaterials can induce DNA damage, genomic instability and epigenetic alteration, which can contribute to the initiation and progression of malignancies and genetic disorders (Ray et al., 2009). Nanomaterials are reported to trigger immune responses, leading to inflammation and potential immune system dysfunction. Nanomaterials that are absorbed into the bloodstream accumulate in various organs, potentially causing toxicity to the organs. Some nanomaterials show the ability to cross the blood–brain barrier, eliciting concerns about their potential neurotoxic effects on the central nervous system (Aljabali et al., 2023; Feng et al., 2015).

The recovery of nanomaterials from wastewater is essential because their presence in effluent streams increases load on treatment systems and potentially harms the ecosystem, aquatic life and microorganisms present in soil. Wastewater streams are not captured effectively by conventional wastewater treatment systems. This leads to the release of nanomaterials into water bodies and affect the quality of drinking water. Therefore, it is high time to do an assessment, treatment and management of nanomaterials in water and wastewater and find out some technologies to treat these nanomaterials.

In the present chapter, we put forward the importance of managing nanomaterial-contaminated effluent, especially graphene oxide (GO) and carbon nanotubes (CNT), by employing recovery methods based on microemulsion and cloud point extraction (CPE) techniques.

L. Dhiman · D. Bhardwaj · V. K. Gahlot · S. K. Singh · P. K. Rai
Centre for Fire, Explosive, and Environment Safety (CFEES) Defence Research & Development Organisation (DRDO), Delhi, India

L. Dhiman · S. Anand (✉)
Department of Environmental Engineering, Delhi Technological University, Delhi, India
e-mail: shalinianand1975@gmail.com

© Capital Publishing Company, New Delhi, India 2025
I. Haq et al. (eds.), *Water and Wastewater*, https://doi.org/10.1007/978-3-031-94253-2_7

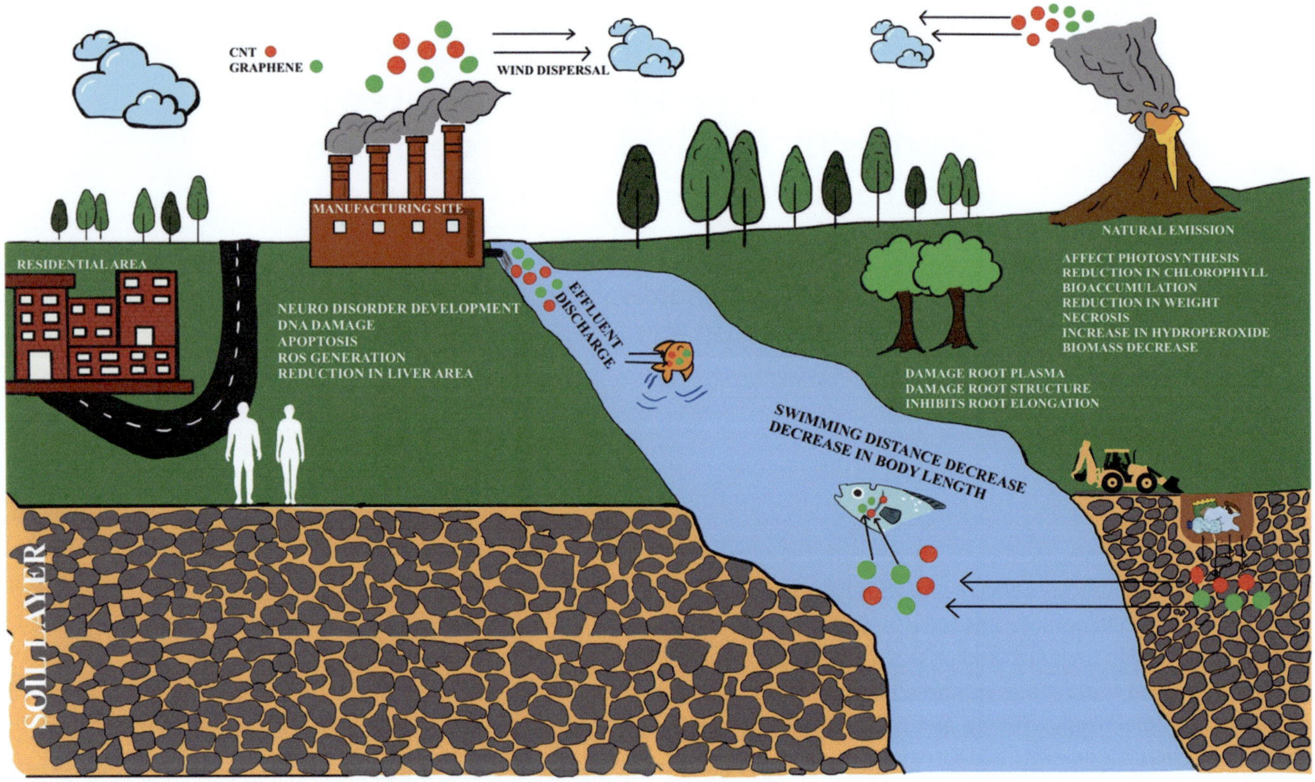

Fig. 1 Nanomaterials synthesis, application and its interaction with the environment

The toxicity of carbon-based nanomaterials (CBNMs) has been elaborated. The current research practices on decontamination of CBNMs from contaminated water sources by different methodologies are reviewed. Lastly, future directions for developing more efficient and selective decontamination methods are highlighted.

2 Nanomaterials and Their Forms

A nanometer is one billionth of a metre, making nanomaterials extremely small and often at the scale of individual molecules or atoms. Many nanomaterials are present with specific properties and potential applications (Anjum et al., 2023). Nanomaterials are categorised according to their unique structure and properties (Salem et al., 2022).

2.1 Metal-Based Nanomaterials

Metal nanoparticles include gold, silver, platinum nanoparticles and many more. Metal oxide nanoparticles include titanium dioxide, zinc oxide and iron oxide nanoparticles used in sunscreens, electronics and environmental remediation. Metal-based quantum dot nanoparticles, including semiconductor nanoparticles, have unique optical properties, for which they are used in displays, imaging and sensors (Yaqoob et al., 2020).

2.2 Carbon-Based Nanomaterials

Carbon-based nanomaterials include graphene, CNT, and fullerene (C60). Graphene is a single layer of carbon atoms arranged in a two-dimensional (2D) hexagonal lattice, known for its excellent electrical and thermal conductivity. CNTs are cylindrical structures made of rolled graphene sheets and used in electronics, composites and drug delivery applications. Fullerenes are spherical carbon molecules, the most well-known being buckminsterfullerene (C60), used in research and potential drug delivery applications (Ma et al., 2023).

2.3 Nanocomposites

Mixtures of nanomaterials with other materials result in enhanced properties. For example, adding nanoparticles to polymers improves their strength, conductivity and other characteristics. For instance, nanowires and nanorods are one-dimensional (1D) nanostructures with high aspect ratios, used in electronics, sensors and energy storage (Colorado et al., 2024).

2.4 Other Types of Nanomaterials

Nanoporous materials with nanoscale pores or voids are used for adsorption, catalysis and separation processes. Nanoporous metals have controlled porosity at nanoscale, and are used in catalysis and filtration applications. Nanoclays are layered silicate nanoparticles used as fillers in polymers to enhance mechanical properties and barrier properties. Liposomes and nanospheres are lipid- or polymer-based nanoparticles used for drug delivery and targeted therapies. Nanoemulsions are emulsions having a size in nanometer range, and are used in pharmaceuticals, cosmetics and food products.

3 Approaches for CBNM Synthesis

The synthesis of CBNMs involves various approaches that yield materials with distinct properties and structures. Top-down and bottom-up are the most used methods to approach nanoscale.

3.1 Top-Down Methods

The top-down approach involves lessening the size of assembly in the direction of nanoscale (Abid et al., 2022).

3.1.1 Synthesis of Graphene Oxide

Top-down approaches for the synthesis of graphene oxide involve exfoliation and oxidation of bulk graphite to produce monolayer or few-layer sheets of graphene with oxygen-containing functional groups. One of the common methods, generally used for the synthesis of graphene oxide is the Hummers' method (Hummers & Offeman, 1958). In Hummers' method, oxidation of graphite is done by using a mixture of concentrated sulphuric acid (H_2SO_4) and nitric acid (HNO_3) at elevated temperatures. This mixture generates strong oxidising agents that break graphite layers and introduce oxygen-containing functional groups. After oxidation, exfoliation was performed to form graphene oxide by adding the mixture of graphite and oxidising acids along with potassium permanganate ($KMnO_4$) and heating. This step leads to the intercalation of oxygen and the formation of epoxy, hydroxyl and carboxyl functional groups on graphene layers, resulting in graphene oxide. After the reaction, the resulting mixture is washed extensively to remove residual acids and manganese oxide. The washed material is typically collected through filtration. The obtained materials were then further functionalized and reduced by reducing oxygen groups and restoring some of the graphene structure. This reduction step is achieved using various reducing agents, such as hydrazine, or by thermal treatment.

3.1.2 Synthesis of Carbon Nanotubes

Synthesis of CNT by top-down approach is done by mechanical exfoliation, laser ablation, etc. CNTs are exfoliated from multiwalled carbon nanotubes (MWNTs) using mechanical forces. Techniques, such as sonication or high-energy milling, are employed to separate individual nanotubes from the bulk material. In the laser ablation method, a laser beam is focused on a carbon target in the presence of a suitable solvent or gas (Gacem et al., 2022). The high-energy laser pulses vapourise the carbon, creating plasma, from which CNTs are formed and collected.

3.2 Bottom-Up Methods

Bottom-up approaches involve building bigger nanomaterials from smaller building blocks or precursors.

3.2.1 Synthesis of Graphene Oxide

Although chemical vapour deposition (CVD) is frequently used for growing graphene, the process can be modified for producing graphene oxide. In this process, hydrocarbon gas is decomposed on a metal catalyst substrate and oxygen-containing gases are introduced to oxidise graphene layers. The oxidised layers then exfoliate to form graphene oxide. Annealing is used to reduce oxygen content and restore some graphene-like properties.

Electrochemical exfoliation is another process where graphite electrodes are subjected to electrochemical processes in an electrolyte (Achee et al., 2018). The electrochemical reactions lead to the exfoliation and oxidation of graphite layers, resulting in the formation of graphene oxide.

3.2.2 Synthesis of Carbon Nanotubes

Synthesis of CNT involves the most prominent CVD method in which the catalytic decomposition of hydrocarbon gases is carried out over a metal catalyst at high temperatures. The metal catalyst particles act as nucleation sites for growth of CNT. The arrangement of carbon atoms forms nanotubes around these catalyst particles. One more method in practice is arc discharge in which CNT can be synthesised by passing a high-current arc through two graphite electrodes in an inert gas atmosphere. The high temperatures generated cause the vapourisation of carbon, which then condenses to form CNT (Vir Singh et al., 2023).

4 Emissions and Toxicity of CBNMs While Transferring to Aqua Sphere

The amount of nanomaterial in water varies widely based on the source of contamination, such as industrial processes and wastewater discharge. According to literature, the estimated concentration of nanomaterials present in wastewater treatment plant and water sources is as described in Tables 1 and 2.

The concentration of nanomaterials present in water confirmed toxic effects on organisms. Graphene oxide is the most toxic form of CBNM tested in mice lungs (Duch et al., 2011), with toxicity in the range of 50–300 mg/L in microbial communities (Ahmed & Rodrigues, 2013). Owing to a wide range of toxicity, it exhibits variations across different levels of organisms, revealing its potential ecological impact. In lower organisms, such as bacteria and algae, graphene oxide disrupts cellular membranes, impact metabolic processes and inhibit growth. In higher organisms, including aquatic organisms, such as fish, and mammals, such as rodents, graphene oxide exposure could lead to oxidative stress, inflammation and organ damage. Understanding graphene oxide toxicity across various organismic levels is crucial for assessing its potential environmental consequences. CNTs have demonstrated toxicity across various organisms because of their unique physicochemical properties. In vitro studies demonstrated that CNTs penetrate cell membranes, leading to oxidative stress, inflammation and DNA damage. In vivo experiments with animal models indicated that CNT exposure could result in pulmonary inflammation, fibrosis and impaired lung function. Moreover, CNTs' potential to translocate and accumulate in vital organs raises concerns about systemic toxicity. Recently in a study on algae, it was discovered that the binary grouping of graphene and funtionalised MWCNT instigated amplified impairment to cells than individual nanomaterials (Das et al., 2023b). Understanding the mechanisms underlying CNT-induced toxicity is essential for assessing their safe use and potential impact on human health and the environment (Francis & Devasena, 2018). Therefore, curtailing the toxic effects of CBNM, and adopting various management strategies for the removal of CBNM from contaminated water is of prime importance.

Table 1 Emissions of CBNMs from wastewater treatment plants

Component	Name of CBNMs	Concentration	References
Effluent	Fullerenes (C_{60})	4 ng/L	Gottschalk et al. (2009)
		1.3–13 ng/L	Gottschalk et al. (2012)
	CNT	3.6–16 ng/L	Gottschalk et al. (2012)
Sewage sludge	Fullerenes (C_{60})	0.09–0.34 mg/kg	Gottschalk et al. (2012)
	CNT	0.12–0.40 mg/kg	Gottschalk et al. (2012)
		3.69–32.66 ng/L	Maurer-Jones et al. (2013)

Notes CBNM: carbon-based nanomaterials; CNT: carbon nanotubes

Table 2 Concentration of nanomaterials in water sources

Component	Name of CBNMs	Concentration	References
Natural water	Fullerenes (C_{60})	0.008–59 ng/L	Zhu et al. (2019)
Surface water	Fullerenes (C_{60})	0.003 ng/L	Gottschalk et al. (2009); Maurer-Jones et al. (2013)
		0.07–0.33 ng/L	Gottschalk and Nowack (2012)
		0.017 ng/L	Gottschalk et al. (2013)
	CNT	0.004 ng/L	Gottschalk et al. (2013)
		0.001–0.8 ng/L	Maurer-Jones et al. (2013)
		0.17–0.56 ng/L	Gottschalk et al. (2012)
Freshwater	Fullerenes (C_{60})	567 ng/L	Tao et al. (2015)
	CNT	0.0025 mg/L	Eckelman et al. (2012)
		0.1–120 mg/L	Freixa et al. (2018)
Sea water	CNT	0.02–0.2 pg/L	Gottschalk et al. (2015)

Notes CBNM: carbon-based nanomaterials; CNT: carbon nanotubes

5 Practices for Recovery of Nanomaterials

Recovery of different types of nanomaterials is done by various methods, such as filtration, centrifugation, precipitation, dialysis and ultrafiltration, evaporation and condensation, chemical precipitation, reverse osmosis, size-exclusion chromatography, magnetic separation, coagulation and flocculation methods. In addition to these, microemulsion and CPE methods for recovery were attempted for a few nanomaterials.

Ultrafiltration is a simple and effective method to separate nanomaterials from wastewater. In this technique, different types of membranes, such as polytetrafluoroethylene (PTFE), polycarbonate track-etched (PCTE) and mixed cellulose ester (MCE) with different pore sizes were used to recover nanomaterials from water (Lee et al., 2020; Zhang et al., 2023). Limitation of pore size and size constraints of nanoparticles reduce filtration efficiency; hence, ultrafiltration is not suitable for all types of nanomaterials (Chu et al., 2022).

Another technique called magnetic separation is effective for nanoparticles having magnetic properties. All nanomaterials do not have magnetic properties; this limits the usage of magnetic separation for the recovery of all nanomaterials (Illés et al., 2014). Coagulation and flocculation methods for the recovery of nanomaterials require more optimisation, as agglomeration occurs in this case, which could be due to the specific properties of nanomaterials (Gregory, 2005). To overcome the above-discussed limitations, chemicals in minute concentrations could be used for the effective recovery of nanomaterials (Bratby, 2013). Chemicals are used in microemulsion and CPE methods.

Of the different studies done so far, 16 studies were conducted for recovery of nanosized metal oxides by CPE technique from surface, potable and sewage water (Table 3). Recovered nanomaterials were characterised by *transmission electron microscopy* (TEM) imaging.

The present research work aimed at the management of CBNMs waste, viz. graphene oxide and CNT, using microemulsion and CPE approaches from nanomaterial-contaminated water samples. The microemulsion technique is based on a minuscule oil–water edge trapping mechanism. The CPE technique considers the innate behaviour of a non-ionic surfactant to assemble into micelles and undergoes clouding above the critical strength of a micelle when heated above the least temperature required for reaction, that is, cloud point temperature (CPT). Once the CPT is achieved, the mixture splits into two diverse phases. The surfactant-laden phase contains nanoparticles entrapped inside the micelles and the surfactant-deprived phase lacks nanoparticles. The surfactant used was non-ionic Triton X-100 in different concentrations, and the salt employed was sodium chloride (NaCl). The collected nanoparticles and supernatant from both methods were characterised by ultraviolet–visible (UV–Vis) spectroscopy, Raman spectroscopy and TEM imaging.

6 Case Study for Recovery of Carbon-Based Nanomaterials via Microemulsion and Cloud Point Extraction Techniques

The adverse effects of nanomaterials' concentration persist in the environment through various sources, such as water, sand and air, seeking more attention for their recovery. Nowadays, because of rapid industrialisation, CBNMs are being synthesised in large quantities and used in various applications, such as semiconductor devices, batteries and many more. CBNMs are discharged into the environment during their synthesis, and after their end use, they are dumped into landfills. The recovery of CBNMs is more complex than that of other nanomaterials, as CBNMs typically have a very small particle size and may form agglomerates or clusters. Dispersing and separating them from a matrix or solvent is challenging. Therefore, for environmental and economic aspects, it is essential to optimise the recovery of graphene oxide and CNT from wastewater.

6.1 Extraction of Nanomaterials Through Microemulsion Method

A microemulsion is a thermodynamically stable and optically transparent dispersion of two immiscible liquids (usually oil and water) stabilised by an emulsifying agent or surfactant. Unlike traditional emulsions, which separate over time into distinct phases, microemulsions remain stable for extended periods because of their unique structure and properties. Tiwari et al. (2015) conducted research, showing that microemulsion technique effectively achieves the extraction of carbon nanotubes, graphene, boron nitride nanotubes, boron nitride nanosheets and zinc oxide nanowires from contaminated water. Notably, Mdlovu et al. (2018) demonstrated the retrieval of copper nanoparticles from waste etchants of printed circuit boards. Myakonkaya et al. (2011) investigated the polymer-mediated recovery of nanoparticles. Finally, Lazim et al. (2010) explored the successful recovery of gold nanoparticles.

The laboratory-based investigation, primarily focusing on the recovery of CBNMs based on microemulsion, represents a straightforward and efficient approach for retrieving nanomaterials from water. Within the microemulsion framework, nanomaterials were separated through the intricacies of microscopic oil–water interfaces. Notably, this technique

Table 3 Recovery of nanomaterials using the cloud point extraction method

S. no	Nanomaterials	Source of nanomaterials	Surfactant used	Recovery rate (%)	References
1	Palladium	Synthetic water	Triton X-100 + Triton X-114	50	Nazar et al. (2011)
2	Copper(II) oxide	Environmental water samples and synthetic water	Triton X-100 + Triton X-114	91	Majedi et al. (2014)
3	Silver	Environmental water samples	Triton X-114	57–116	Liu et al. (2009a)
4	Silver	Antibacterial products, environmental water samples and synthetic water	Triton X-114	71.7–103	Chao et al. (2011)
5	Silver	Environmental water samples	Triton X-114	79–122	Yang et al. (2016)
6	Silver	Environmental water samples and synthetic water	Triton X-114	82–100	Bahadir et al. (2018)
7	Gold	Synthetic water	Triton X-114	92–97	Liu et al. (2009b)
8	Gold	Synthetic water	Triton X-100 + Triton X-114	52	Nazar et al. (2011)
9	Gold	Environmental water samples and synthetic water	Triton X-114	91–103	Hartmann and Schuster (2013)
10	Gold	Soil extract and synthetic water	Triton X-114	80–112	El Hadri and Hackley (2017)
11	Gold	Environmental water samples and synthetic water	Triton X-114	78.8–109.8	Mandyla et al. (2017)
12	Gold	Environmental water samples and synthetic water	Triton X-114	94–100	Bahadir et al. (2018)
13	ZnO	Environmental water samples	Triton X-114	87.3	Majedi et al. (2012)
14	Fe_3O_4	Synthetic water	Triton X-114	90–97	Liu et al. (2009b)
15	Titanium oxide (TiO_2)	Environmental water samples and synthetic water	Triton X-114	99	López-Mayan et al. (2019)
16	SiO_2	Environmental water samples and synthetic water	Triton X-114	107	Yang et al. (2017)

capitalises on the formation of micelles at ambient temperatures, a phenomenon induced by the introduction of paraffin-heavy oil. These micelles effectively encapsulate nanomaterials within the hydrophobic confines of micellar environment. The micelles formed by oil and water phase were separated by filtration or ejection, and the separated materials were subjected to ethanol washing to remove oily phase from the surface of nanomaterials. However, this methodology leads to a low recovery rate. Figure 2 illustrates the mechanism underlying microemulsion technique, in which the nanomaterials are entrapped within the emulsion formed at oil and water interface.

Raman spectra and TEM imaging of graphene oxide and CNT in contaminated water and after treatment are depicted in Fig. 3. The spectral analysis shows the absence of nanomaterials in post-treatment water, indicating successful recovery through the microemulsion process. In addition, the morphological characteristics of nanomaterials as shown through TEM imaging evidently reinforce the notion that microemulsion treatment could be used for the recovery of nanomaterials in their original shape and size.

6.2 Extraction of Nanomaterials Through Cloud Point Extraction Method

Cloud point extraction is a sample preparation technique used to extract nanomaterials from aqueous solutions. CPE is based on the principle that non-ionic surfactants form micelles above their CPT. Micelles are aggregates of surfactant molecules that have a hydrophobic core and a hydrophilic shell. The hydrophobic core of micelles can solubilise hydrophobic compounds, such as nanomaterials. Numerous prior investigations have explored the recovery of nanometal oxides using CPE technique. However, retrieval of CBNMs through this method has not been accomplished. CPE, also

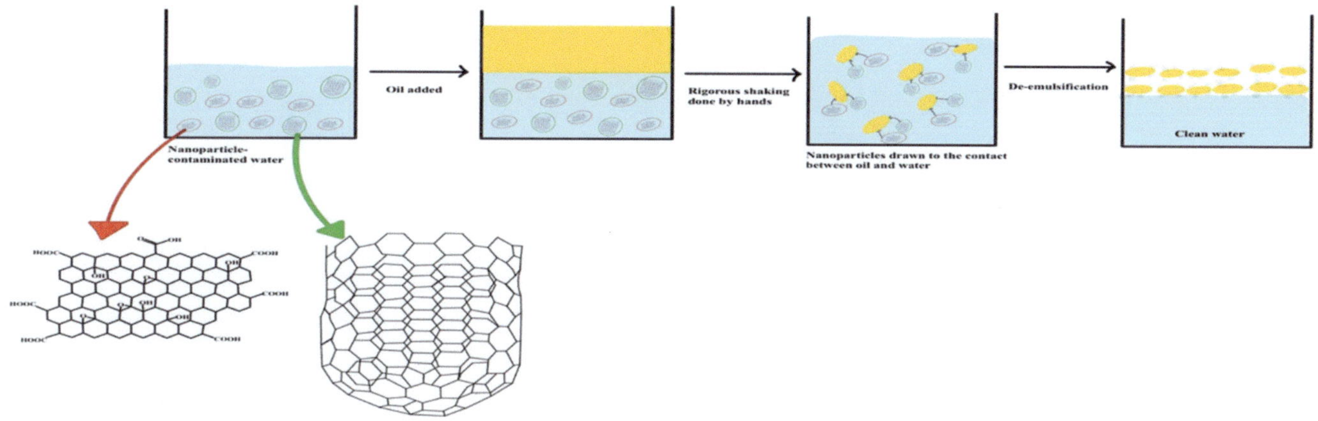

Fig. 2 Schematic of recovery by microemulsion technique

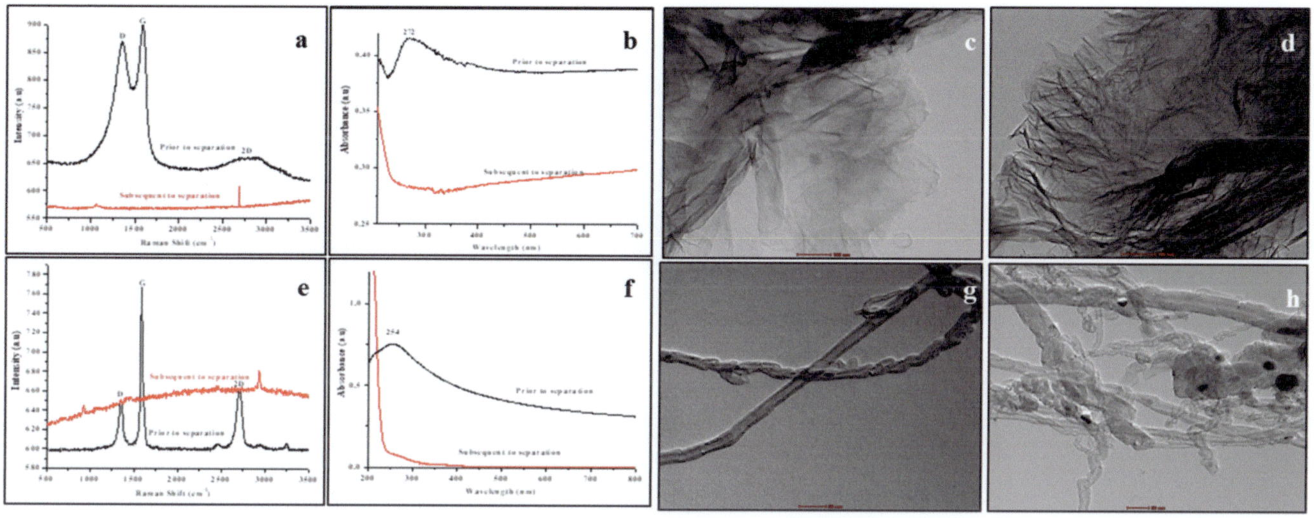

Fig. 3 Representation of **a** Raman spectra of GO, **b** UV–Vis spectra of GO, **c** TEM image of GO before microemulsion treatment, **d** TEM image of GO after microemulsion treatment, **e** Raman spectra of CNT, **f** UV–Vis spectra of CNT, **g** TEM image of CNT before microemulsion treatment, **h** TEM image of CNT after microemulsion treatment

known as cloud point preconcentration or cloud point solubilisation, is a separation technique used to concentrate and extract target analytes from a solution using a surfactant phase transition. It is particularly useful for extracting trace amounts of organic and inorganic compounds from complex matrices, such as aqueous solutions. In Fig. 4, the intricate mechanism of CPE is elucidated, shedding light on the intricate processes that govern this technique's effectiveness in separating and recovering target analytes from complex solutions.

Within the context of CPE, Triton X-100 offers a versatile, uncomplicated and economically viable pathway for the efficient isolation of nanomaterials from aqueous phases. Remarkably, this method yields a recovery rate exceeding 97% for nanomaterials. Notably, both UV–Vis and Raman analyses reveal an absence of significant peaks corresponding to nanomaterials, underscoring the efficacy of the process. Furthermore, the nanomaterials exhibit no discernible morphological alterations, as confirmed by TEM spectroscopy.

Figure 5 represents the Raman spectra of both graphene oxide and CNT from contaminated water before and after the treatment. The spectral analysis indicates that the absence of nanomaterials in post-treatment water demonstrates successful recovery by CPE process. In addition, the morphological characteristics of nanomaterials represented through TEM imaging specifically strengthen the concept that CPE process effectively preserves the original shape and size of nanomaterials.

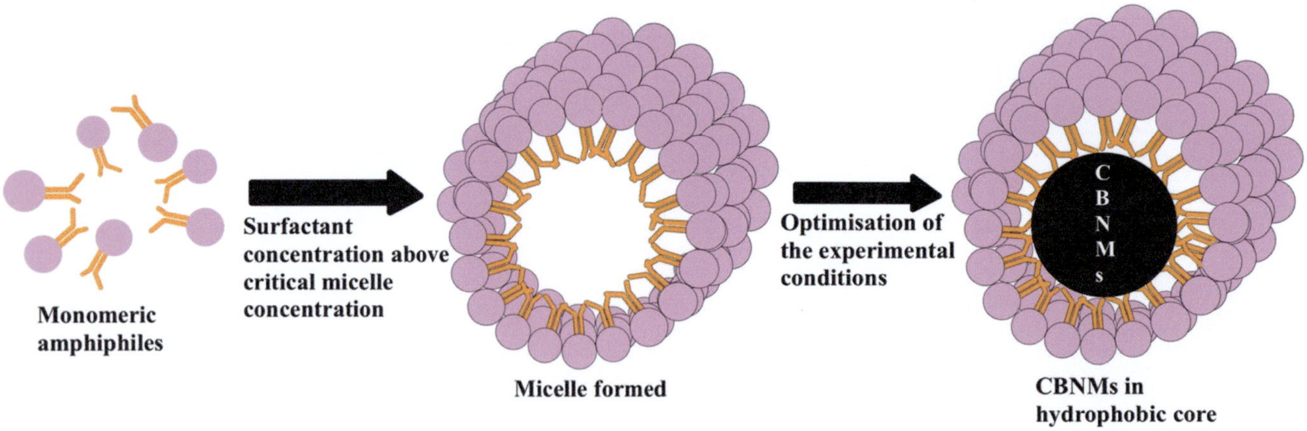

Fig. 4 Mechanism of recovery by cloud point extraction

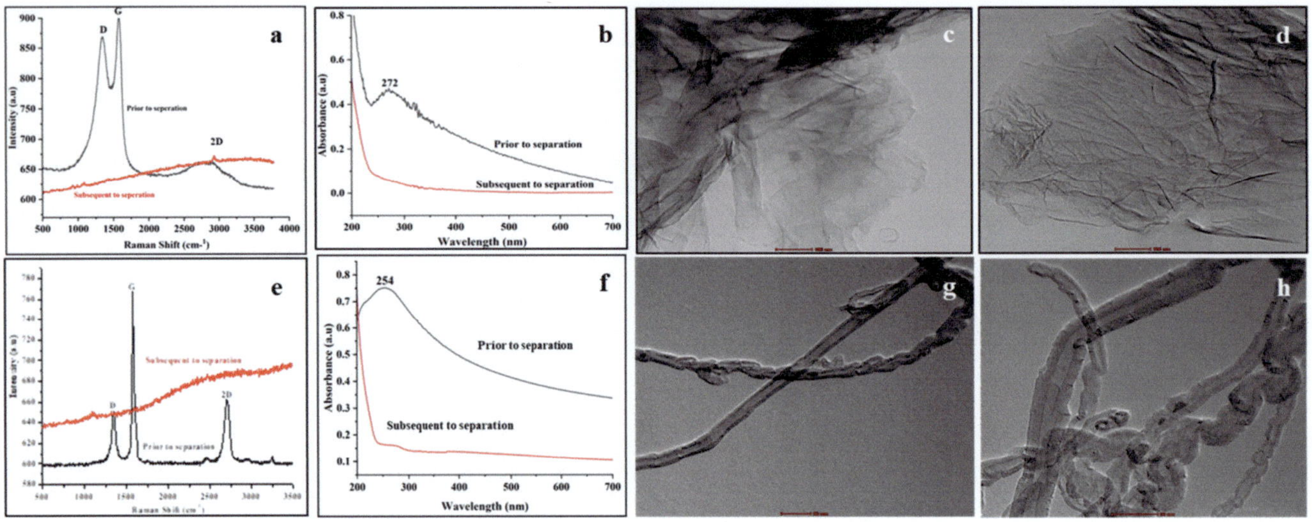

Fig. 5 Representation of **a** Raman spectra of GO, **b** UV–Vis spectra of GO, **c** TEM image of GO before CPE treatment, **d** TEM image of GO after CPE treatment, **e** Raman spectra of CNT, **f** UV–Vis spectra of CNT, **g** TEM image of CNT before CPE treatment, **h** TEM image of CNT after CPE treatment

7 Conclusions

A confined number of conventional methods available for the recovery of nanomaterials are associated with drawbacks in terms of their efficiency. Both CPE and microemulsion techniques were harnessed for the extraction of CBNMs from contaminated water sources. Within the CPE approach, a phenomenon occurring at a specific temperature, inclusion of a surfactant into the solution initiates the formation of micelles. These micelles efficiently encapsulate the intended analytes, subsequently eliminating them from the solution. The segregated micellar phase is then isolated, facilitating the recovery of analytes. CPE technique is more efficient than microemulsion as it can extract more of the targeted nanomaterials from contaminated sources. Notably, CPE technique exhibits a heightened capability in the retrieval of CBNMs, compared to microemulsion process.

References

Abid, N., Khan, A. M., Shujait, S., Chaudhary, K., Ikram, M., Imran, M., Haider, J., Khan, M., Khan, Q., & Maqbool, M. (2022). Synthesis of nanomaterials using various top-down and bottom-up approaches, influencing factors, advantages, and disadvantages: A review. *Advances in Colloid and Interface Science, 300*, Article 102597. https://doi.org/10.1016/j.cis.2021.102597

Achee, T. C., Sun, W., Hope, J. T., Quitzau, S. G., Sweeney, C. B., Shah, S. A., Habib, T., & Green, M. J. (2018). High-yield scalable graphene nanosheet production from compressed graphite using electrochemical exfoliation. *Scientific Reports, 8*, 14525. https://doi.org/10.1038/s41598-018-32741-3

Ahmed, F., & Rodrigues, D. F. (2013). Investigation of acute effects of graphene oxide on wastewater microbial community: A case study. *Journal of Hazardous Materials, 256*, 33–39. https://doi.org/10.1016/j.jhazmat.2013.03.064

Aljabali, A. A., Obeid, M. A., Bashatwah, R. M., Serrano-Aroca, Á., Mishra, V., Mishra, Y., El-Tanani, M., Hromić-Jahjefendić, A., Kapoor, D. N., Goyal, R., & Naikoo, G. A. (2023). Nanomaterials and their impact on the immune system. *International Journal of Molecular Sciences, 24*(3), 2008. https://doi.org/10.3390/ijms24032008

Anjum, A., Garg, R., Kashif, M., & Eddy, N. O. (2023). Nano-scale innovations in packaging: Properties, types, and applications of nanomaterials for the future. *Food Chemistry Advances, 3*, Article 100560. https://doi.org/10.1016/j.focha.2023.100560

Azim, M. M., Arifutzzaman, A., Saidur, R., Khandaker, M. U., & Bradley, D. A. (2022). Recent progress in emerging hybrid nanomaterials towards the energy storage and heat transfer applications: A review. *Journal of Molecular Liquids, 360*, Article 119443. https://doi.org/10.1016/j.enmm.2022.100719

Bahadir, Z., Torrent, L., Hidalgo, M., & Marguí, E. (2018). Simultaneous determination of silver and gold nanoparticles by cloud point extraction and total reflection X-ray fluorescence analysis. *Spectrochimica Acta Part B: Atomic Spectroscopy, 149*, 22–29. https://doi.org/10.1016/j.sab.2018.07.016

Bratby, J. (2013). *Coagulation and flocculation in water and wastewater treatment*, 3rd Ed. IWA Publishing, London.

Chao, J. B., Liu, J. F., Yu, S. J., Feng, Y. D., Tan, Z. Q., Liu, R., & Yin, Y. G. (2011). Speciation analysis of silver nanoparticles and silver ions in antibacterial products and environmental waters via cloud point extraction-based separation. *Analytical Chemistry, 83*(17), 6875–6882. https://doi.org/10.1021/ac201086a

Chu, L. K., Wickramasinghe, S. R., Qian, X., & Zydney, A. L. (2022). Retention and fouling during nanoparticle filtration: Implications for membrane purification of biotherapeutics. *Membranes, 12*(3), 299. https://doi.org/10.3390/membranes12030299

Colorado, H. A., Gutierrez-Velasquez, E. I., Gil, L. D., & de Camargo, I. L. (2024). Exploring the advantages and applications of nanocomposites produced via vat photopolymerization in additive manufacturing: A review. *Advanced Composites and Hybrid Materials, 7*, 1. https://doi.org/10.1007/s42114-023-00808-z

Das, S., Giri, S., Jose, S. A., Pulimi, M., Anand, S., Chandrasekaran, N., Pramod, K. R., & Mukherjee, A. (2023a). Comparative toxicity assessment of individual, binary and ternary mixtures of SiO_2, Fe_3O_4, and ZnO nanoparticles in freshwater microalgae, *Scenedesmus obliquus*: Exploring the role of dissolved ions. *Comparative Biochemistry and Physiology Part c: Toxicology & Pharmacology, 273*, Article 109718. https://doi.org/10.1016/j.cbpc.2023.109718

Das, S., Giri, S., Wadhwa, G. Pulimi, M., Anand, S., Chandrasekaran, N., Johari, S. A., Rai, P. K., & Mukherjee, A. (2023b). Comparative ecotoxicity of graphene, functionalized multi-walled CNTs, and their mixture in freshwater microalgae, Scenedesmus obliquus: Analyzing the role of oxidative stress. *Environmental Science and Pollution Research (ESPR), 30*, 70246–70259. https://doi.org/10.1007/s11356-023-27367-6

Duch, M. C., Budinger, G. S., Liang, Y. T., Soberanes, S., Urich, D., Chiarella, S. E., Campochiaro, L. A., Gonzalez, A., Chandel, N. S., Hersam, M. C., & Mutlu, G. M. (2011). Minimizing oxidation and stable nanoscale dispersion improves the biocompatibility of graphene in the lung. *Nano Letters, 11*(12), 5201–5207. https://doi.org/10.1021/nl202515a

Eckelman, M. J., Mauter, M. S., Isaacs, J. A., & Elimelech, M. (2012). New perspectives on nanomaterial aquatic ecotoxicity: Production impacts exceed direct exposure impacts for carbon nanotoubes. *Environmental Science & Technology, 46*(5), 2902–2910. https://doi.org/10.1021/es203409a

El Hadri, H., & Hackley, V. A. (2017). Investigation of cloud point extraction for the analysis of metallic nanoparticles in a soil matrix. *Environmental Science: Nano, 4*(1), 105–116. https://doi.org/10.1039/C6EN00322B

Feng, X., Chen, A., Zhang, Y., Wang, J., Shao, L., & Wei, L. (2015). Central nervous system toxicity of metallic nanoparticles. *International Journal of Nanomedicine, 10*, 4321–4340. https://doi.org/10.2147/IJN.S78308

Francis, A. P., & Devasena, T. (2018). Toxicity of carbon nanotubes: A review. *Toxicology and Industrial Health, 34*(3), 200–210. https://doi.org/10.1177/0748233717747472

Freixa, A., Acuña, V., Sanchís, J., Farré, M., Barceló, D., & Sabater, S. (2018). Ecotoxicological effects of carbon based nanomaterials in aquatic organisms. *Science of the Total Environment, 619*, 328–337. https://doi.org/10.1016/j.scitotenv.2017.11.095

Gacem, A., Modi, S., Yadav, V. K., Islam, S., Patel, A., Dawane, V., Jameel, M., Inwati, G. K., Piplode, S., Solanki, V. S., & Basnet, A. (2022). Recent advances in methods for synthesis of carbon nanotubes and carbon nanocomposite and their emerging applications: A descriptive review. *Journal of Nanomaterials, 7238602*, 1–16. https://doi.org/10.1155/2022/7238602

Giese, B., Klaessig, F., Park, B., Kaegi, R., Steinfeldt, M., Wigger, H., von Gleich, A., & Gottschalk, F. (2018). Risks, release and concentrations of engineered nanomaterial in the environment. *Scientific Reports, 8*(1), 1565. https://doi.org/10.1038/s41598-018-19275-4

Gómez-López P., Puente-Santiago A., Castro-Beltrán A., do Nascimento L. A., Balu A. M., Luque R., & Alvarado-Beltrán C. G. (2020). Nanomaterials and catalysis for green chemistry. *Current Opinion in Green and Sustainable Chemistry, 24*, 48–55. https://doi.org/10.1016/j.cogsc.2020.03.001

Gong, S., & Cheng, W. (2017). One-dimensional nanomaterials for soft electronics. *Advanced Electronic Materials, 3*(3), 1600314. https://doi.org/10.1002/aelm.201600314

Gottschalk, F., Lassen, C., Kjoelholt, J., Christensen, F., & Nowack, B. (2015). Modeling flows and concentrations of nine engineered nanomaterials in the Danish environment. *International Journal of Environmental Research and Public Health, 12*(5), 5581–5602. https://doi.org/10.3390/ijerph120505581

Gottschalk, F., & Nowack, B. (2012). Modeling the environmental release and exposure of engineered nanomaterials. In: T. Puzyn & J. Leszczynski (Eds.), *Towards efficient designing of safe nanomaterials* (Chap. 12, pp. 284–313). Royal Society of Chemistry. https://doi.org/10.1039/9781849735476

Gottschalk, F., Sonderer, T., Scholz, R. W., & Nowack, B. (2009). Modeled environmental concentrations of engineered nanomaterials (TiO_2, ZnO, Ag, CNT, fullerenes) for different regions. *Environmental Science & Technology, 43*(24), 9216–9222. https://doi.org/10.1021/es9015553

Gottschalk, F., Sun, T., & Nowack, B. (2013). Environmental concentrations of engineered nanomaterials: Review of modeling and analytical studies. *Environmental Pollution, 181*, 287–300. https://doi.org/10.1016/j.envpol.2013.06.003

Gregory, J. (2005). *Particles in water: Properties and processes.* CRC Press. https://doi.org/10.1201/9780203508459

Gupta, V., Mohapatra, S., Mishra, H., Farooq, U., Kumar, K., Ansari, M. J., Aldawsari, M. F., Alalaiwe, A. S., Mirza, M. A., & Iqbal, Z. (2022). Nanotechnology in cosmetics and cosmeceuticals—A review of latest advancements. *Gels, 8*(3), 173. https://doi.org/10.3390/gels8030173

Hartmann, G., & Schuster, M. (2013). Species selective preconcentration and quantification of gold nanoparticles using cloud point extraction and electrothermal atomic absorption spectrometry. *Analytica Chimica Acta, 761*, 27–33. https://doi.org/10.1016/j.aca.2012.11.050

Hummers, W. S., Jr., & Offeman, R. E. (1958). Preparation of graphitic oxide. *Journal of the American Chemical Society, 80*(6), 1339. https://doi.org/10.1021/ja01539a017

Illés, E., Szekeres, M., Kupcsik, E., Tóth, I. Y., Farkas, K., Jedlovszky-Hajdú, A., & Tombácz, E. (2014). PEGylation of surfacted magnetite core–shell nanoparticles for biomedical application. *Colloids and Surfaces a: Physicochemical and Engineering Aspects, 460*, 429–440. https://doi.org/10.1016/j.colsurfa.2014.01.043

Imran, A. B., & Susan, M. A. B. H. (2022). Natural fiber-reinforced nanocomposites in automotive industry. In H. Song, T. A. Nguyen, G. Yasin, N. B. Singh, & R. K. Gupta (Eds.), *Nanotechnology in the automotive industry* (pp. 85–103). Elsevier BV. https://doi.org/10.1016/B978-0-323-90524-4.00005-0

Lazim, A. M., Eastoe, J., Bradley, M., Trickett, K., Mohamed, A., & Rogers, S. E. (2010). Recovery of gold nanoparticles using pH-sensitive microgels. *Soft Matter, 6*(9), 2050–2055. https://doi.org/10.1039/C002511A

Lee, H., Segets, D., Süß, S., Peukert, W., Chen, S. C., & Pui, D. Y. (2020). Effects of filter structure, flow velocity, particle concentration and fouling on the retention efficiency of ultrafiltration for sub-20 nm gold nanoparticles. *Separation and Purification Technology, 241*, Article 116689. https://doi.org/10.1016/j.seppur.2020.116689

Liu, J. F., Chao, J. B., Liu, R., Tan, Z. Q., Yin, Y. G., Wu, Y., & Jiang, G. B. (2009a). Cloud point extraction as an advantageous preconcentration approach for analysis of trace silver nanoparticles in environmental waters. *Analytical Chemistry, 15*, 6496–6502. https://doi.org/10.1021/ac900918ea

Liu, J. F., Liu, R., Yin, Y. G., & Jiang, G. B. (2009b). Triton X-114 based cloud point extraction: A thermoreversible approach for separation/concentration and dispersion of nanomaterials in the aqueous phase. *Chemical Communications, 12*, 1514–1516. https://doi.org/10.1039/B821124Hb

López-Mayan, J., Barciela-Alonso, M. C., Domínguez-González, M. R., Pena-Vazquez, E., & Bermejo-Barrera, P. (2020). Cloud point extraction and ICP-MS for titanium speciation in water samples. *Microchemical Journal, 152*, Article 104264. https://doi.org/10.1016/j.microc.2019.104264

Ma, J., Wang, G., Ding, X., Wang, F., Zhu, C., & Rong, Y. (2023). Carbon-based nanomaterials as drug delivery agents for colorectal cancer: Clinical preface to colorectal cancer citing their markers and existing theranostic approaches. *ACS Omega, 8*, 10656–10668. https://doi.org/10.1021/acsomega.2c06242

Majedi, S. M., Kelly, B. C., & Lee, H. K. (2014). Toward a robust analytical method for separating trace levels of nano-materials in natural waters: Cloud point extraction of nano-copper (II) oxide. *Environmental Science and Pollution Research, 21*(20), 11811–11822. https://doi.org/10.1007/s11356-013-2381-7

Majedi, S. M., Lee, H. K., & Kelly, B. C. (2012). Chemometric analytical approach for the cloud point extraction and inductively coupled plasma mass spectrometric determination of zinc oxide nanoparticles in water samples. *Analytical Chemistry, 84*(15), 6546–6552. https://doi.org/10.1021/ac300833t

Mandyla, S. P., Tsogas, G. Z., Vlessidis, A. G., & Giokas, D. L. (2017). Determination of gold nanoparticles in environmental water samples by second-order optical scattering using dithiotreitol-functionalized CdS quantum dots after cloud point extraction. *Journal of Hazardous Materials, 323*, 67–74. https://doi.org/10.1016/j.jhazmat.2016.03.039

Maurer-Jones, M. A., Gunsolus, I. L., Murphy, C. J., & Haynes, C. L. (2013). Toxicity of engineered nanoparticles in the environment. *Analytical Chemistry, 85*(6), 3036–3049. https://doi.org/10.1021/ac303636s

Mdlovu, N. V., Chiang, C. L., Lin, K. S., & Jeng, R. C. (2018). Recycling copper nanoparticles from printed circuit board waste etchants via a microemulsion process. *Journal of Cleaner Production, 185*, 781–796. https://doi.org/10.1016/j.jclepro.2018.03.087

Myakonkaya, O., Eastoe, J., Mutch, K. J., & Grillo, I. (2011). Polymer-induced recovery of nanoparticles from microemulsions. *Physical Chemistry Chemical Physics, 13*(8), 3059–3063. https://doi.org/10.1039/C0CP00736F

Nazar, M. F., Shah, S. S., Eastoe, J., Khan, A. M., & Shah, A. (2011). Separation and recycling of nanoparticles using cloud point extraction with non-ionic surfactant mixtures. *Journal of Colloid and Interface Science, 363*(2), 490–496. https://doi.org/10.1016/j.jcis.2011.07.070

Ray, P. C., Yu, H., & Fu, P. P. (2009). Toxicity and environmental risks of nanomaterials: Challenges and future needs. *Journal of Environmental Science and Health Part C, 27*(1), 1–35. https://doi.org/10.1080/10590500802708267

Salem, S. S., Hammad, E. N., Mohamed, A. A., & El-Dougdoug, W. (2022). A comprehensive review of nanomaterials: Types, synthesis, characterization, and applications. *Biointerface Research in Applied Chemistry, 13*, 41. https://doi.org/10.33263/BRIAC131.041

Shah, M. A., Pirzada, B. M., Price, G., Shibiru, A. L., & Qurashi, A. (2022). Applications of nanotechnology in smart textile industry: A critical review. *Journal of Advanced Research, 38*, 55–75. https://doi.org/10.1016/j.jare.2022.01.008

Shilar, F. A., Ganachari, S. V., & Patil, V. B. (2022). Advancement of nano-based construction materials—A review. *Construction and Building Materials, 359*, Article 129535. https://doi.org/10.1016/j.conbuildmat.2022.129535

Tao, X., Yu, Y., Fortner, J. D., He, Y., Chen, Y., & Hughes, J. B. (2015). Effects of aqueous stable fullerene nanocrystal (nC60) on *Scenedesmus obliquus*: Evaluation of the sub-lethal photosynthetic responses and inhibition mechanism. *Chemosphere, 122*, 162–167. https://doi.org/10.1016/j.chemosphere.2014.11.035

Tiwari, B., Zhang, D., Winslow, D., Lee, C. H., Hao, B., & Yap, Y. K. (2015). A simple and universal technique to extract one- and two-dimensional nanomaterials from contaminated water. *ACS Applied Materials & Interfaces, 7*(47), 26108–26116. https://doi.org/10.1021/acsami.5b07542

Vir Singh, M., Kumar Tiwari, A., & Gupta, R. (2023). Catalytic chemical vapor deposition methodology for carbon nanotubes synthesis. *Chemistry Select, 8*, Article e202204715. https://doi.org/10.1002/slct.202204715

Xiong, F., Huang, S., & Gu, N. (2018). Magnetic nanoparticles: Recent developments in drug delivery system. *Drug Development and Industrial Pharmacy, 44*(5), 697–706. https://doi.org/10.1080/03639045.2017.1421961

Yang, Y., Luo, L., Li, H. P., Wang, Q., Yang, Z. G., & Long, C. L. (2016). Separation and determination of silver nanoparticle in environmental water and the UV-induced photochemical transformations study of AgNPs by cloud point extraction combined ICP-MS. *Talanta, 161*, 342–349. https://doi.org/10.1016/j.talanta.2016.08.052

Yang, Y., Reed, R., Schoepf, J., Hristovski, K., Herckes, P., & Westerhoff, P. (2017). Prospecting nanomaterials in aqueous environments by cloud-point extraction coupled with transmission electron microscopy. *Science of the Total Environment, 584*, 515–522. https://doi.org/10.1016/j.scitotenv.2017.01.059

Yaqoob, A. A., Ahmad, H., Parveen, T., Ahmad, A., Oves, M., Ismail, I. M., Qari, H. A., Umar, K. M., & Ibrahim, M. N. (2020). Recent advances in metal decorated nanomaterials and their various biological applications: A review. *Frontiers in Chemistry, 8*, 2020. https://doi.org/10.3389/fchem.2020.00341

Zhang, J., Poon, W., & Chen, S. C. (2023). Experimental and theoretical retentions of sub-10 nm colloidal nanoparticles by large-pore ultrafiltration membranes in isopropanol and water. *Journal of Membrane Science, 674*, Article 121487. https://doi.org/10.1016/j.memsci.2023.121487

Zhou, X., Jin, W., & Ma, J. (2023). Lung inflammation perturbation by engineered nanoparticles. *Frontiers in Bioengineering and Biotechnology, 11*, 1199230. https://doi.org/10.3389/fbioe.2023.1199230

Zhu, Y., Liu, X., Hu, Y., Wang, R., Chen, M., Wu, J., Wang, Y., Kang, S., Sun, Y., & Zhu, M. (2019). Behavior, remediation effect and toxicity of nanomaterials in water environments. *Environmental Research, 174*, 54–60. https://doi.org/10.1016/j.envres.2019.04.014

Isolation and Identification of Indigenous Bacteria for Microbial Bioremediation Study

Varsha Thapa and Dechen C. Tsering

1 Introduction

Sikkim, surrounded by the Tibet Plateau in the North, West Bengal in the South, Kingdom of Bhutan and the Chumbi Valley of Tibet in the East and Nepal in the West, is one of the smallest states in India. The rise in global population has led to the rise in pharmaceutical industries for the production of various pharmaceutical products and a wide group of medications. The exemption from income tax and excise duty, subsidy in investment of plant and machinery, and reimbursements in insurance plans offered under the North-East Industrial and Investment Promotion Policy (NEIIPP), 2007, have attracted many pharmaceutical companies to invest in the state of Sikkim. Many pharmaceutical companies are located near the riverside of Rangpo, Ranikhola in the East district that connects the river Teesta, making this district over populated than the South, West and North districts of Sikkim (Adhikari, 2018).

The pharmaceutical industries generate toxic pollutants, known as pharmaceutical wastewater. These toxic pollutants are treated by effluent treatment plants (ETPs) that involve physico-chemical and biological processes to reduce pollution before its release in the nearby water bodies. The untreated pharmaceutical wastewater serves as a potential hotspot of various microorganisms, such as bacteria, that have the ability to degrade chemical contaminants present in the untreated water, thus reducing the toxicity of pollutants (Kotwani et al., 2021).

Microbial bioremediation could give rise to low-cost, efficient and environment-friendly bacterial strains for biodegradation of contaminants present in pharmaceutical wastewater (Singh et al., 2020). Isolation, screening and identification of indigenous bacteria from pharmaceutical effluents of the Sikkim region have not been studied; till date, no bioremediation studies using indigenous bacteria have been conducted for biodegradation of toxic pollutants present in pharmaceutical effluents into non-toxic form. This biological tool of microbes could help us understand the process of bioremediation, and bacterial isolates could work as an inoculum for reduction of various physico-chemical parameters, resulting in decrease of pollutant load of pharmaceutical wastewater.

2 Materials and Method

2.1 Study Design and Sampling

A field study-based design. The untreated wastewater samples (inlet effluent) were collected from 10 pharmaceutical companies of East Sikkim from April to September 2022 using thoroughly cleaned and sterile 500-mL Pyrex bottles, each at the discharge point of common effluent treatment plant where oil and water were separated. All samples from each pharmaceutical unit were collected in triplicate and transported to the laboratory under refrigerated conditions and stored at 2–8 °C (Hingurao & Nerurkar, 2018). All the experiments were performed within the shortest time possible after sample collection to avoid any kind of chemical transformation that might occur due to prolonged storage of samples (Ibrahim et al., 2020).

2.2 Isolation of Bacterial Isolates

The untreated wastewater samples (inlet effluent) were diluted serially (10^{-1}–10^{-6}) in physiological saline solution (0.9% NaCl w/v). Solution from appropriate dilution, 0.1 mL, was spread plated on nutrient agar (NA) inside a laminar floor hood and incubated at 37 °C for 48–72 h under inverted stationary conditions. Plates were observed for bacterial growth. For

V. Thapa (✉) · D. C. Tsering
Department of Microbiology, Sikkim Manipal Institute of Medical Sciences, Sikkim Manipal University, Tadong, Gangtok, Sikkim, India
e-mail: varshathapa90@gmail.com

© Capital Publishing Company, New Delhi, India 2025
I. Haq et al. (eds.), *Water and Wastewater*, https://doi.org/10.1007/978-3-031-94253-2_8

strain purification, streaking inoculation was done repeatedly to obtain a single colony. The pure bacterial strain was inoculated onto nutrient agar slants and incubated for 1–2 days and preserved in a refrigerator at 4 °C. All pure isolates were preserved at 80 °C in nutrient broth containing 25% glycerol (Li & Zhang, 2018).

2.3 Morphological Identification and Biochemical Characterisation of Bacterial Isolates

The morphological characteristics of pure bacterial isolates (size, shape, opacity, elevation, margin and pigmentation) were observed by Gram staining. The biochemical characterisation of bacteria was done by the conventional method, wherein the Gram-negative bacteria was identified based on catalase test, citrate utilisation, indole test, methyl red test, urease test, Voges Proskauer test, sugar fermentation test and H_2S production. Gram-positive isolates were also differentiated by Gram staining, colonial characteristics, catalase test and coagulase tests. All these tests were performed according to standard methods (Gowri et al., 2020).

2.4 Physico-Chemical Analysis of Untreated Pharmaceutical Effluent Samples

The untreated pharmaceutical wastewater was checked for the following parameters: pH using a pH meter, 5-day biochemical oxygen demand (BOD), chemical oxygen demand (COD) by the Open Reflux method, total dissolved solids (TDS) and total suspended solids (TSS); this was done as per the standard methods of the American Public Health Association (APHA, 2017).

2.5 Bioremediation Treatment Study of Untreated Wastewater with Indigenous Bacterial Isolates

2.5.1 Preparation of Inoculum

In 100-mL nutrient broth media, 0.1 mL of each identified bacterial isolate was inoculated in inoculated flasks, which were kept on a rotary shaker at 150 rpm for 24 h at 35 °C. The biomass of actively growing cells was prepared for studying the biodegradation efficiency of identified bacterial isolates. After the incubation period, the active microbial cultures were centrifuged at 7,000 rpm for 10 min at 4 °C. The supernatant was discarded, and the pellet was washed thrice with sterile deionised water. This was done for all identified isolates, and the pellet of each bacterial isolate was stored for further experiments (Al-Wasify et al., 2017; Ibrahim et al., 2020).

2.5.2 Optimization of Various Parameters (Inoculum Size, Temperature, Agitation Speed and Incubation Time) for Their Roles in Biodegradation Study

The bacterial inoculum of all isolated bacteria was prepared as mentioned above. Pellets were re-suspended for the time of the experiment. Untreated pharmaceutical wastewater, 500 mL, was taken in an Erlenmeyer flask, and bacterial inoculum of 3%, 6% and 9% (v/v) of all bacterial isolates was assessed for analysing biodegradation. The incubation period for all flasks with untreated pharmaceutical wastewater inoculated with the above-mentioned percentage of bacterial isolates was maintained at 35, 37 and 40° C in a shaking incubator at 120, 150 and 200 rpm. The samples were withdrawn after 24, 48 and 72 h. Controls without bacterial inoculation were routinely assessed (Srivastava et al., 2022). All the experiments were done in triplicate. The physico-chemical parameters, such as pH, BOD, COD, TDS and TSS, were measured as per the standard methods of APHA (2017).

3 Results and Discussion

3.1 Isolation and Morphological Characterisation of Bacterial Isolates

The following 22 different types of pharmaceutical inlet isolates (PII) were examined in the untreated wastewater of 10 pharmaceutical units: three isolates from sample 1: PII1, PII2 and PII3; four isolates from sample 2: PII4, PII5, PII6 and PII7; two isolates from sample 3, namely: PII8 and PII9; three isolates from sample 4: PII10, PII11 and PII12; one isolate from sample 5: PII13; three isolates from sample 6: PII14, PII15 and PII16; two isolates from sample 7: PII17 and PII18; two isolates from sample 8, namely PII19 and PII20; one isolate from sample 9: PII21; and one isolate from sample 10: PII22. The morphological and colony characteristics of all isolated bacteria are presented in Table 1. Among 22 bacterial isolates, isolated from the untreated inlet wastewater, Gram-positive character was exhibited by three isolates and these were mainly cocci and bacilli. The remaining 19 bacterial isolates were Gram-negative bacilli. The microscopic characteristics of potential bacterial isolates are shown in Fig. 1.

PII are the isolates insulated from untreated inlet samples of 10 pharmaceutical units. The conventional biochemical method resulted in the characterisation of 22 bacterial isolates as listed in Table 2. Isolates PII1, PII3 and PII13 were characterised as *Staphylococcus aureus* sp.; P2, PII8 and PII18 isolates were characterised as *Pseudomonas putida* sp.; PII17 and PII19 isolates were *Escherichia coli* sp.; PII4, PII7 and PII13 isolates were identified as *Klebsiella pneumoniae* sp.; PII10, PII15 and PII20 isolates were identified as *Klebsiella*

Table 1 The colony characteristics and cell morphology of all isolated bacterial cultures

Bacterial isolates	Margin of colony	Elevation of colony	Opacity/ Translucency	Surface of colony	Form of colony	Colour (in NA media)	Gram stain
PII1	Entire	Convex	Opaque	Smooth	Circular	Yellow	Gram-positive cocci
PII2	Entire	Convex	Opaque	Smooth	Circular	Pale green	Gram-negative rod
PII3	Entire	Convex	Opaque	Smooth	Circular	Yellow	Gram-positive cocci
PII4	Entire	Slightly raised	Opaque/ translucent	Mucoid	Circular	White	Gram-negative rod
PII5	Entire	Sightly raised	Opaque	Smooth	Circular	White	Gram-negative rod
PII6	Entire	Convex	Opaque	Smooth	Circular	Blue-green	Gram-negative rod
PII7	Entire	Slightly raised	Opaque/ translucent	Mucoid	Circular	White	Gram-negative rod
PII8	Entire	Convex	Opaque	Smooth	Circular	Pale green	Gram-negative rod
PII9	Entire	Raised	Opaque	Smooth	Circular	Cream	Gram-negative rod
PII10	Entire	Raised	Opaque/ translucent	Mucoid	Circular	Cream	Gram-negative rod
PII11	Entire	Raised	Opaque	Smooth	Circular	Cream	Gram-negative rod
PII12	Entire	Sightly raised	Opaque	Smooth	Circular	White	Gram- negative rod
PII13	Entire	Slightly raised	Opaque/ translucent	Mucoid	Circular	White	Gram-negative rod
PII14	Entire	Convex	Opaque	Smooth	Circular	Yellow	Gram-negative cocci
PII15	Entire	Raised	Opaque/ translucent	Mucoid	Circular	Cream	Gram-negative rod
PII16	Entire	Raised	Opaque	Smooth	Circular	White/Greyish	Gram-positive rod
PII17	Entire	Slightly raised	Opaque/ translucent	Smooth	Circular	White/Greyish	Gram-negative rod
PII18	Entire	Convex	Opaque	Smooth	Circular	Pale green	Gram-negative rod
PII19	Entire	Slightly raised	Opaque/ translucent	Smooth	Circular	White/Greyish	Gram-negative rod
PII20	Entire	Raised	Opaque/ translucent	Mucoid	Circular	Cream	Gram-negative rod
PII21	Entire	Convex	Opaque	Smooth	Circular	White	Gram-negative rod
PII22	Entire	Convex	Opaque	Smooth	Circular	Blue-green	Gram-negative rod

oxytoca sp.; PII5 and PII12 isolates were characterised as *Acinetobacter baumannii*; PII9 and PII11 isolates were characterised as *Enterobacter cloacae*; PII22 and PII6 isolates were characterised as *Pseudomonas aeruginosa*; PII21 isolate was identified as *Serratia liquefaciens*; and PII16 isolate was identified as *Bacillus subtilis* sp.

3.2 Physico-Chemical Analysis of Untreated Pharmaceutical Wastewater Samples

The physico-chemical parameters are a significant foundation for developing, manufacturing and maintaining the stability of a pharmaceutical product (Hamaidi-Chergui & Errahmani,

Fig. 1 **a** Bacterial isolates in nutrient agar media. **b** Biochemical characterisation of isolated bacterial isolates

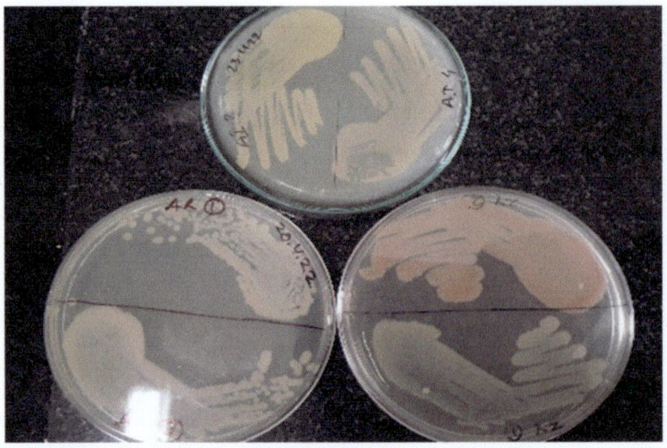

(a)

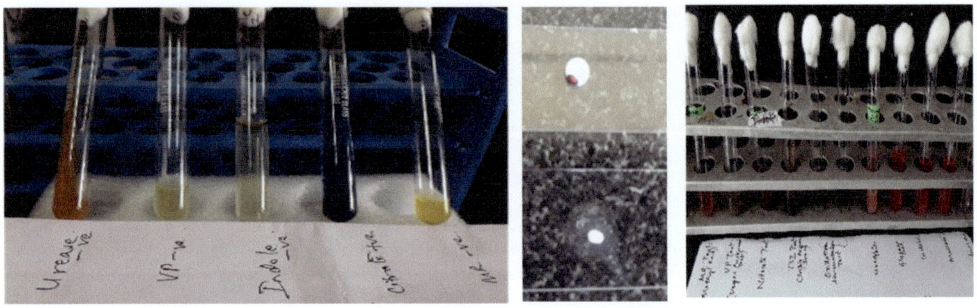

(b)

2019). The study on characterisation and physico-chemical analysis of the inlet of wastewater treatment plant revealed that the wastewater contains a high amount of BOD, COD, TDS and TSS. The study also showed the efficiency of the treatment plant in controlling the organic and microbial pollutant load of wastewater (Dehghani et al., 2018).

The results of physico-chemical parameters from inlet of 10 pharmaceutical units are listed in Table 3. The results show an increase in all physico-chemical parameters, as the raw samples contained high number of organic substances and suspended materials that have not been processed for treatment in effluent treatment plants as shown in Fig. 2.

The untreated pharmaceutical wastewater showed an increase in all physico-chemical parameters because of the presence of biodegradable organic matter, such as prescription drugs, antibiotics, non-prescription drugs, personal care products and other pollutants (Rana et al., 2014).

3.3 Bioremediation Treatment Study of Untreated Wastewater with Indigenous Bacterial Isolates

In this study, 22 different types of bacterial isolates isolated from the inlet of 10 pharmaceutical units were used to check the biodegradation efficiency of bacteria for microbial bioremediation. Different parameters, such as percentage of inoculum, temperature, time and agitation speed required for bacterial growth, incubated in the untreated wastewater sample UPWW4 to check the biodegradation percentage of potential bacteria in reducing pollutant load. The UPWW4 sample was taken because it had the highest pollutant load among the other wastewater samples.

From 22 different bacterial isolates, three bacterial isolates, namely *P. aeruginosa, A. baumannii* and *S. liquefaciens*, confirmed by conventional biochemical method, showed potential results in reducing BOD, COD, TDS and TSS of untreated wastewater samples. The results of microbial bioremediation of the three most potential bacterial isolates are listed in Table 4.

The results of this study showed that isolate PII22 identified as *P. aeruginosa* could reduce BOD by 67.7%, COD by 72.67%, TDS by 56.66% and TSS by 43.67%. A study conducted by Wahab and Ali (2020) reported the ability of *P. aeruginosa* in reducing the pollutant load of the Tikrit Wastewater Treatment Plant. This bacterium could reduce COD by 82.5%, thereby increasing the efficiency of wastewater treatment plants. A study reported by Ratna and Kumar (2022) stated the potential of *P. aeruginosa* SRRBL1 in detoxification and degradation of distillery wastewater. In 20% (v/v) distillery wastewater, incubation of 5% inoculum of *P. aeruginosa* SRRBL1 at 37 °C, 120 rpm for 72 h reduced BOD by 61.76%, TSS by 61.32% and COD by 53.53%. 20% (v/v) distillery wastewater, incubation of 5% inoculum of *P.*

Table 2 The biochemical test results for bacteria isolated from untreated pharmaceutical wastewater samples (Kamatham et al., 2024)

Biochemical tests	Bacterial isolates									
	PII1, PII3, PII14	P2, PII8, PII18	PII17, PII19	PII4, PII7, PII13	PII10, PII15, PII20	PII5, PII12	PII9, PII11	PII22, PII6	PII21	PII16
Citrate	+	–	–	+	+	+	+	+	+	+
Nitrate	+	–	+	+	+	–	+	+	+	+
Indole	–	–	+	–	+	–	–	–	–	–
TSI	A/A	K/K	A/A	A/K	A/A	K/K	K/A	A/A	A/K	K/A
Catalase	+	+	+	+	+	+	+	+	+	+
Oxidase	–	+	–	–	–	–	–	+	–	+
MR	+	–	+	–	–	–	–	–	–	–
VP	+	–	–	+	+	–	+	–	+	+
Lactose	+	–	+	+	+	+	+	–	+	+
Fructose	–	–	+	+	+	–	–	–	+	–
Sucrose	+	–	–	+	+	–	+	–	+	+
Maltose	–	–	–	+	+	–	+	–	+	–
Bacterial identity	*Staphylococcus aureus*	*Pseudomonas putida*	*Escherichia coli*	*Klebsiella pneumoniae*	*Klebsiella oxytoca*	*Acinetobacter baumannii*	*Enterobacter cloacae*	*Pseudononas aeruginosa*	*Serratia liquefaciens*	*Bacillus subtilis*

Note PII: pharmaceutical inlet isolate

Table 3 The physico-chemical analysis of untreated pharmaceutical wastewater samples

Samples	Physico-chemical parameters			
	BOD (mg/L)	COD (mg/L)	TDS (mg/L)	TSS (mg/L)
UPWW1	44	640	1,316	228
UPWW2	59.15	546	1,066	240
UPWW3	34	256	800	210
UPWW4	164	660	1,516	290
UPWW5	32.60	372	1,066	218
UPWW6	42.96	532	1,510	180
UPWW7	52.56	404	1,288	270
UPWW8	45.43	470	1,245	233
UPWW9	61.23	650	785	177
UPWW10	38.05	575	677	332

Notes UPWW: untreated pharmaceutical wastewater; COD: chemical oxygen demand; BOD: biochemical oxygen demand; TDS: total dissolved solids; TSS: total suspended solids

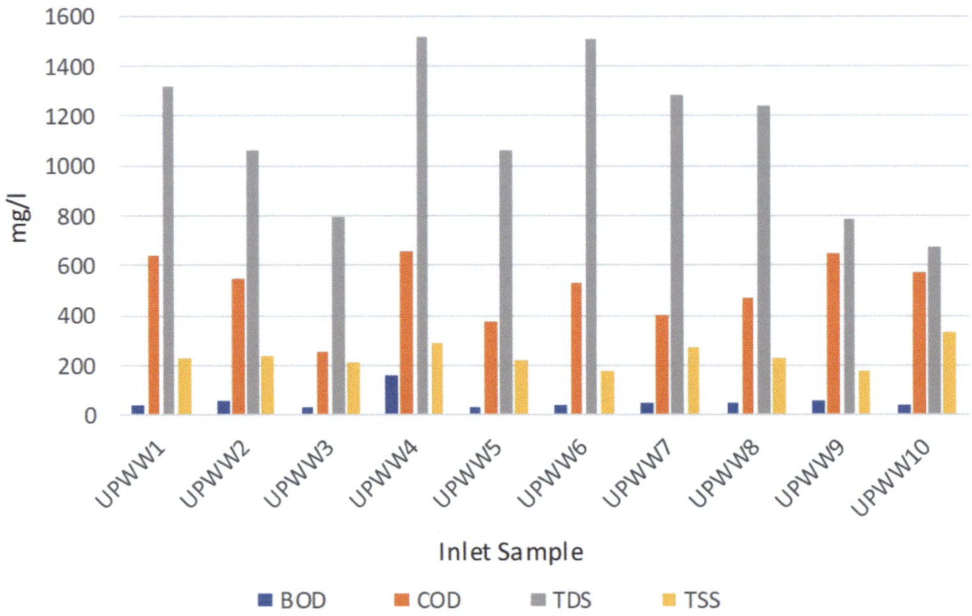

Fig. 2 The physico-chemical parameters of untreated (inlet) pharmaceutical wastewater

aeruginosa SRRBL1 at 37 °C, 120 rpm for 72 h reduced BOD by 61.76%, TSS by 61.32% and COD by 53.53%.

The bacterial isolate PII5 identified as *A. baumannii* showed reduction of BOD by 61.43%, COD by 68.21%, TDS by 54.57% and TSS by 59.44% on treating this bacterium with untreated inlet pharmaceutical wastewater. In a study conducted by Kharat et al. (2019), involving sewage treatment by *A. baumanii* strain VESASC-ASC, showed 70% reduction in BOD and 99.26% reduction in COD at 35–37 °C in 8–12 days. The sewage wastewater inoculated with the above-mentioned bacterial strain *A. baumanii* strain VESASC-ASC showed promising results, compared with the wastewater inoculated with the extracted extracellular polymeric substances (EPS) of the *A. baumanii* strain VESASC-ASC. Shaker et al. (2022) studied the potential of *A. baumanii* in bacterial consortium in the biodegradation of widely used fluoroquinolone (FQs) antibiotics. Removal of FQs from wastewater using traditional methods has been ineffective and is a serious concern for various health risks to all living organisms.

The third promising isolate for biodegradation study is the isolate PII21 identified as *S. liquefaciens*. It could reduce BOD by 53.47%, COD by 62.91%, TDS by 61.33% and TSS by 48.17% on addition of 9% of the bacterial inoculum at 37 °C at 120 rpm in 72 h. The biodegradation study of the above three mentioned bacterial isolates with the untreated wastewater and the control study in which there is only wastewater and no bacterial isolate are shown in

Table 4 Bioremediation treatment study of untreated pharmaceutical wastewater samples with potential bacterial isolates

Sample	Bioremediation parameters						Physico-chemical parameters			
	Bacterial isolate	Inoculum size (%)	Temperature (°C)	rpm	Time (hr)		BOD (mg/L)	COD (mg/L)	TDS (mg/L)	TSS (mg/L)
UPWW4	*Pseudomonas aeruginosa*	9	37	120	72	Sample	52.97	180.37	657.03	163.35
						Control	164	660	1,516	290
						Reduction (%)	67.7	72.67	56.66	43.67
UPWW4	*Acinetobacter baumannii*	9	37	120	72	Sample	63.25	209.81	688.71	117.62
						Control	164	660	1,516	290
						Reduction (%)	61.43	68.21	54.57	59.44
UPWW4	*Serratia liquefaciens*	9	37	120	72	Sample	76.30	244.79	586.23	150.30
						Control	164	660	1,516	290
						Reduction (%)	53.47	62.91	61.33	48.17

Fig. 3. Gupta and Thakur (2015) studied the biodegradation of wastewater using an indigenous bacterial strain *Serratia* sp. ISTVKR1 isolated from sewage sludge. The results showed the removal of complex organic contaminants and removal of COD. The study revealed the capacity of microbial strains and antagonistic interactions between different strains present in the sewage wastewater that could result as a potential bio-degrader. The isolation, screening and application of indigenous bacterial strain could add to the efficiency and enhance the removal of contaminants in treatment plants. Bhatt et al. (2022) reported the application of *Serratia* sp. isolated from wastewater treatment plants in degradation of tetracycline, an antibiotic used in treating various diseases. *Serratia* sp. could degrade 89.5% of tetracycline in 48 h under optimised conditions of pH 6.0, 20 mg/L of tetracycline concentration at 36 °C using 100-μL inoculum. This study demonstrated the use of indigenous microbial strains in degradation of tetracycline, an antibiotic produced at a large scale in various pharmaceutical industries.

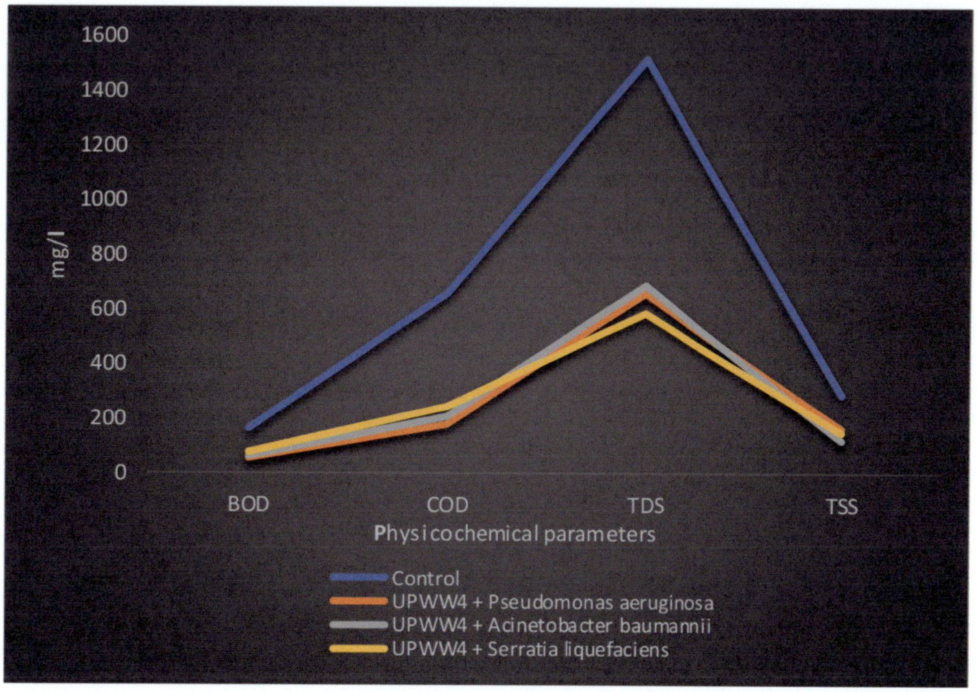

Fig. 3 Bioremediation treatment of UPWW4 samples with potential bacterial isolates *P. aeruginosa, A. baumannii* and *S. liquefaciens*. UPWW4: untreated pharmaceutical wastewater; BOD: biochemical oxygen demand; COD: chemical oxygen demand; TDS: total dissolved solids; TSS: total suspended solids

4 Conclusion

In this study, 22 different types of bacterial isolates were isolated from untreated wastewater (inlet) of 10 pharmaceutical units in East Sikkim. Different parameters, such as time, temperature, rpm and inoculum percentage of bacterial isolates incubated with the inlet sample, were optimised to achieve maximum degradation of organic pollutants present in wastewater. Among the 22 isolates, three bacterial isolates, namely *P. aeruginosa, A. baumannii* and *S. liquefaciens*, as identified by conventional biochemical method showed maximum bioremediation potential for pharmaceutical wastewater treatment. Isolate *Pseudomonas aeruginosa* showed maximum BOD reduction of 67.7% and COD was reduced by 72.67%. Isolate *A. baumannii* could reduce BOD by 61.43% and COD by 68.21%, followed by *S. liquefaciens*, which could reduce BOD by 53.47% and COD by 62.91%. The maximum percentage of TDS reduction was observed on inoculation of the sample with the *S. liquefaciens* isolate, followed by *A. baumannii* that could reduce TSS by 59.44%. The biodegradation potential of the above three bacterial isolates was achieved by inoculating the untreated inlet sample with 9% of bacterial inoculum at 37 °C, 120 rpm for 72 h. The above-mentioned three bacterial isolates could enhance the process of biological treatment installed in the effluent treatment plants of East Sikkim, enabling sufficient microbial growth and biodegradation of pollutants at a faster rate. This study on bioremediation following the concept of bioaugmentation could be helpful for designing of bioreactors for pharmaceutical wastewater treatment that could decrease pollutant load for release of safe and clean water into the rivers of Sikkim.

Acknowledgements The authors express sincere gratitude to the Department of Microbiology, Sikkim Manipal Institute of Medical Sciences (SMIMS), Sikkim Manipal University (SMU), Gangtok, Sikkim, for providing laboratory facilities and resources that played a crucial role in the execution of this research. The author extends gratitude to supervisor, Dr. Dechen C. Tsering, professor, Department of Microbiology, Sikkim Manipal Institute of Medical Sciences, Sikkim Manipal University, Tadong, Gangtok, Sikkim, for her guidance, invaluable insights and continuous encouragement throughout the research process. The authors thank Dr. Gopal Pradhan, Member Secretary, State Pollution Control Board (SPCB), Sikkim, for allowing us to conduct this research at the SPCB, Forest Secretariat, C-Block, Deorali, Gangtok, Sikkim. The authors also thank B.K. Chettri, Principal Chief Scientific Officer, SPCB, Sikkim, for his thoughtful insights and support throughout this work. The authors are grateful to K.Z. Bhutia, Senior Scientific Officer, SPCB, Gangtok, for her guidance and support. The authors are thankful to Mrs. Kusum Gurung, Joint Director, SPCB and Mr. Ranjan Rai, Executive Environmental Engineer (EEE) for their guidance and encouragement. The authors also acknowledge Mr. Gaychung Lepcha, Junior Scientific Staff (JSA), SPCB and Meloom Zangmu Bhutia, Laboratory Assistant (LA), SPCB for their support and cooperation during this work. Further, the authors thank Ritesh Gurung, Scientist D, Central Pollution Control Board (CPCB), Ministry of Environment, Forest and Climate Change, Govt. of India, for being a scientific advisor in this study.

Author Contributions V.T.: Review of literature, preparation of manuscript, conception and design of the study, data collection and preparation of the first draft of the manuscript. D.C.T.: Scientific advisor, contributed to the acquisition and revision of the manuscript.

References

Adhikari, D. (2018). A study on quality of work life in pharmaceuticals company in Sikkim. *International Journal of Science and Research (IJSR), 7*(9), 43–44. https://doi.org/10.21275/ART20191444

Al-Wasify, R. S., Ali, M. N., & Hamed, S. R. (2017). Biodegradation of dairy wastewater using bacterial and fungal local isolates. *Water Science and Technology, 76*(11), 3094–3100. https://doi.org/10.2166/wst.2017.481

American Public Health Association (APHA). (2017). *Standard methods for the examination of water and wastewater* (23rd ed.). APHA.

Bhatt, P., Jeon, C. H., & Kim, W. (2022). Tetracycline bioremediation using the novel *Serratia marcescens* strain WW1 isolated from a wastewater treatment plant. *Chemosphere, 298*, Article 134344. https://doi.org/10.1016/j.chemosphere.2022.134344

Dehghani, M., Delbar Saf, A., Vosoughi, A., Tebbenouri, G., & Ghazanfari Hadi, Z. (2018). Effectiveness of the mindfulness-acceptance-commitment-based approach on athletic performance and sports competition anxiety: A randomized clinical trial. *Electronic Physician, 10*(5), 6749–6755. https://doi.org/10.19082/6749

Gowri, A. K., Margaret, J. K., Vasanthy, M., Ravindran, B., Nguyen-Tri, P., Ngo, H. H., Bui, X. T., Nguyen, X. H., Nguyen, D. D., Chang, S. W., & Chandran, T. (2020). Evaluation of bioremediation competence of indigenous bacterial strains isolated from fabric dyeing effluent. *Bioresource Technology Reports, 11*, Article 100536. https://doi.org/10.1016/j.biteb.2020.100536

Gupta, A., & Thakur, I. S. (2015). Biodegradation of wastewater organic contaminants using *Serratia* sp. ISTVKR1 isolated from sewage sludge. *Biochemical Engineering Journal, 102*, 115–124. https://doi.org/10.1016/j.bej.2015.02.007

Hamaidi-Chergui, F., & Errahmani, M. B. (2019). Water quality and physicochemical parameters of outgoing waters in a pharmaceutical plant. *Applied Water Science, 9*, 165. https://doi.org/10.1007/s13201-019-1046-1

Hingurao, K., & Nerurkar, A. (2018). A novel methylotrophic bacterial consortium for treatment of industrial effluents. *Applied Biochemistry and Biotechnology, 185*(3), 691–704. https://doi.org/10.1007/s12010-017-2680-4

Ibrahim, S., El-Liethy, M. A., Elwakeel, K. Z., Hasan, M. A. E. G., Al Zanaty, A. M., & Kamel, M. M. (2020). Role of identified bacterial consortium in treatment of Quhafa Wastewater Treatment Plant influent in Fayuom. *Egypt. Environmental Monitoring and Assessment, 192*(3), 161. https://doi.org/10.1007/s10661-020-8105-9

Kamatham, S., Seeralan, M., Sekar, U., & Kuppusamy, S. (2024). Antimicrobial resistance profiling of bacteria isolated from wastewater and samples of pharmaceutical industries in South India. *Infection, Genetics and Evolution, 125*, Article 105670. https://doi.org/10.1016/j.meegid.2024.105670

Kharat, A., Ganger, S., & Patil, S. (2019). Extraction of exopolysaccharide from *Acinetobacter baumannii* strain VESASC-ASC and its application in sewage treatment. *International Journal of Emerging Technologies and Innovative Research, 6*(5), 621–629. http://www.jetir.org/papers/JETIR1905H98.pdf

Kotwani, A., Joshi, J., & Kaloni, D. (2021). Pharmaceutical effluent: A critical link in the interconnected ecosystem promoting antimicrobial resistance. *Environmental Science and Pollution Research, 28*(25), 32111–32124. https://doi.org/10.1007/s11356-021-14178-w

Li, Y., & Zhang, K. (2018). Pilot scale treatment of polluted surface waters using membrane-aerated biofilm reactor (MABR). *Biotechnology & Biotechnological Equipment, 32*(2), 376–386. https://doi.org/10.1080/13102818.2017.1399826

Rana, R. S., Singh, P., Singh, R., & Gupta, S. (2014). Assessment of physico-chemical pollutants in pharmaceutical industrial wastewater of Pharmacity, Selaqui, Dehradun. *International Journal of Research in Chemistry and Environment, 4*(2), 136–142. https://doi.org/10.1007/s13201-014-0225-3

Ratna, S., & Kumar, R. (2022). Production of di-rhamnolipid with simultaneous distillery wastewater degradation and detoxification by newly isolated *Pseudomonas aeruginosa SRRBL1. Journal of Cleaner Production, 336*, Article 130429. https://doi.org/10.1016/j.jclepro.2022.130429

Shaker, R. A. E., Nagy, Y. I., Adly, M. E., Khattab, R. A., & Ragab, Y. M. (2022). *Acinetobacter baumannii, Klebsiella pneumoniae and Elizabethkingia miricola* isolated from wastewater have biodegradable activity against fluoroquinolone. *World Journal of Microbiol Biotechnol, 38*(11), 187. https://doi.org/10.1007/s11274-022-03367-5

Singh, P., Singh, V. K., Singh, R., Borthakur, A., Madhav, S., Ahamad A., Kumar, A., Pal, D. B., Tiwary, D., & Mishra, P. K. (2020). Bioremediation: A sustainable approach for management of environmental contaminants. *Abatement of Environmental Pollutants: Trends and Strategies,* 1–23. https://doi.org/10.1016/B978-0-12-818095-2.00001-1

Srivastava, A., Dangi, L. K., Kumar, S., & Rani, R. (2022). Microbial decolorization of Reactive Black 5 dye by *Bacillus albus* DD1 isolated from textile water effluent: Kinetic, thermodynamics & decolorization. *Heliyon., 8*(2), Article e08834. https://doi.org/10.1016/j.heliyon.2022.e08834

Wahab, Z. A. A., & Ali, A. M. H. A. (2020). Treatment of wastewater in the city of Tikrit using bacteria *Pseudomonas* spp. *AIP Conference Proceedings., 2213*(1), Article 020233. https://doi.org/10.1063/5.0000452

Management and Treatment Technologies for Swine Wastewater: A Global Perspective

Monish Goswami and Ajay Kalamdhad

1 Introduction

It was anticipated that the long-term trajectory of China's per capita meat consumption would resume by 2023, as the impact of African swine fever (ASF) on domestic pig meat pricing would diminish (FAO, 2019). Pig meat, consequently, accounts for one-third of the overall increase in meat consumption during the projection period. China would account for 70% of the increase in pig meat consumption between the reference period and 2030. (FAO, 2021). It is anticipated that the global per capita meat consumption will increase by 0.3 percent to 35.4 kg of retail weight equivalent by 2030, from the current level. More than half of this increase is attributable to the increased per capita consumption of poultry meat (FAO, 2021).

Global pig production has increased by more than 3.5 times in the past 40 years, resulting in a higher concentration of animals per farm, sometimes reaching thousands of animals. Substantial quantities of waste, such as wastewater and manure, are produced as the demand for pork meat increases. Pig waste has been employed in the past as soil conditioners and fertilisers. Pig waste is acknowledged as a substantial contributor to global environmental pollution. The transmission of viruses and diseases in pigs is facilitated by the mismanagement and insufficient disposal of pig waste. However, the intensification of cattle rearing has led to severe environmental damage, such as health risks, atmospheric emissions, surface and groundwater contamination, odour dispersion and soil degradation (Sanchez & Leakey, 1997).

Pig manure contains a significant amount of unstabilised organic materials and a high concentration of NH_3-N (Doháнyos et al., 1997). In communities with pig farms, particularly in industrial settings, disease outbreaks are extensively documented. Individuals employed on industrialised pig farms have been linked to outbreaks of methicillin-resistant Staphylococcus aureus (MRSA), which have been attributed to the use of potent antibiotics. Pig farms are also susceptible to transmitting additional diseases, including Salmonella, Toxoplasma and Campylobacter. These diseases can be prevented by implementing appropriate safety measures, including the use of face masks, hand-washing and the covering of open wounds when handling pigs (Turner & Burton, 1997). Pig rearing is essential for the sustainable farm economy of impoverished communities in the northeastern states of India, where piggery waste management studies are scarce.

Vietnamese pig farmers utilise pig manure for fish production due to the unique qualities of these wastes, which frequently result in higher prices (Vu et al., 2007). Anaerobic ponds are frequently employed in Thailand and neighbouring countries to purify piggery wastewater prior to its discharge into water resources. This procedure entails the separation of manure from wastewater. A conventional treatment process for swine wastewater treatment is illustrated in Fig. 1.

2 Global Quantification of Pig Meat

Meat production and consumption are expanding at a faster pace; in 2018, they increased by 47% (109 million tonnes) in comparison to 2000. All in all, pigs, chicken and cattle comprise 90% of the most widely accounted meat for global production. Despite the fact that the proportion of pig meat in the global meat market decreased from 38% in 2000 to 35% in 2018, it still accounts for a total volume of 120,881 thousand tonnes (FAO, 2019). China is the world's largest consumer of pig meat, consuming approximately 45% of the total. Table 1 illustrates a comparison with India.

The data presented above demonstrates that pork is the most commonly consumed meat on Earth. The demand for

M. Goswami (✉) · A. Kalamdhad
Indian Institute of Technology Guwahati, Guwahati, Assam, India
e-mail: monishxgoswami@gmail.com

© Capital Publishing Company, New Delhi, India 2025

I. Haq et al. (eds.), *Water and Wastewater*, https://doi.org/10.1007/978-3-031-94253-2_9

Fig. 1 Conventional treatment process

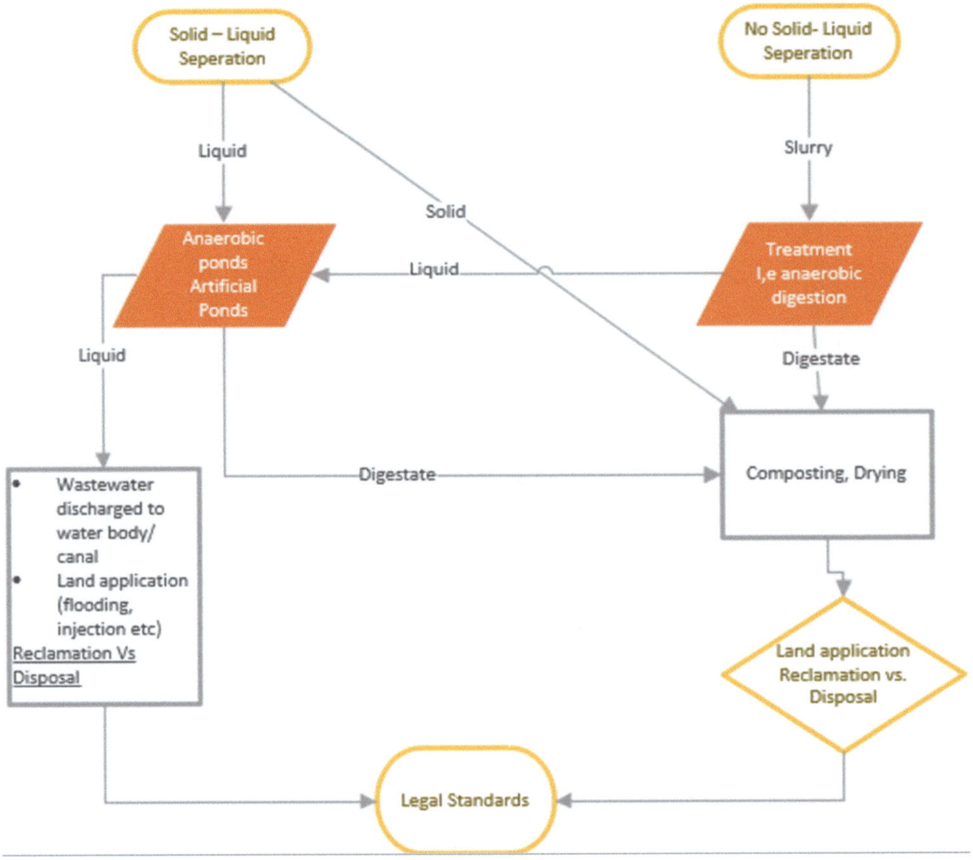

Table 1 World meat production in comparison to India

Country	2018				Total	Meat	(×1000 tonnes)	
	Piggery	Broiler chicken	Cow meat	Others	2000	2010	2015	2018
Global	120,881	114,267	67,354	39,894	233,441	294,619	325,427	342,396
America	22,334	48,363	32,460	4,848	73,793	92,453	99,425	108,005
Asia	66,726	40,036	14,664	22,257	91,355	123,059	138,850	143,684
Europe	296,65	18,649	10,618	4,913	51,305	56,592	61,009	63,846
Africa	1,588	5,748	6,700	6,136	11,585	16,639	19,191	20,171
China	54,984	14,579	5,811	12,783	62,058	80,899	87,651	88,156
(82.4% of Asia)								
India	296	3,591	948	2620	4,449	6,089	7,117	7,454

pig meat is driven by the growing human population. The inevitable consequences of an expanding population and output are a greater carbon footprint and increased piggery waste. The magnitude of the subject 'pigs' is readily apparent from the aforementioned data, so it is imperative to gain a deeper understanding of the subject. Oklahoma State University defines 'swine' as a comprehensive family of animals that generate pork. Nevertheless, the term 'pig' is reserved for the entire swine family, regardless of whether it refers to young pigs or adult hogs. A collective noun for female pigs is 'sows', while a singular noun for male swine is 'boars'.

Annually, the average sow can produce eight to 12 piglets in a single or double litter. Nursery pigs are farm pigs that are less than 23 years old. Pigs that weigh between 23 and 109 kg are classified as finishing pigs. Hogs are adult pigs that weigh a minimum of 109 kg. Mature hogs can weigh up to 227 kg. The pig population in Punjab and Haryana is over 1.5 lakh together. Assam is the state that produces the most pigs in India, with a pig population exceeding 21 lakhs. In the northeast region of India, the total production of pigs exceeds 42 lakhs, with nearly half being raised there (Department of Animal Husbandry and Dairying, 2019).

Table 2 Pig classification and its waste characterisation

Growth phase of pig	Total solids (TS)	Volatile solids (VS) kg/day	COD	Nitrogen	Phosphorus	Potassium	Moisture content	*Moisture %*
Pig 12.5 kg	4.8	4.01	4.4	0.41	0.068	0.16	48	90
Grower 70 kg	5.6	4.5	4.7	4.7	0.76	2.03	560	90
Fertile sow 200 kg	0.5	0.45	0.45	0.032	0.009	0.022	11	90
Lactation sow 192 kg	1.14	1.05	1.09	0.086	0.025	0.054	25	90
Boar 200 kg	0.38	0.34	0.27	0.028	0.011	0.018	8.4	90

Source ASAE (2005)

As per the 20th Livestock Census of India, 2019, the total pig population in the remaining regions of India, except Punjab and Haryana, exceeded 89 lakhs. The north-eastern region of India consumes approximately 70% of the country's pork production in Assam, with an annual consumption of 11.07 crore kg. In Assam, pork is consumed by up to 30% of the population, or 1.03 crore individuals, each year. Nevertheless, the state's annual pork production is 1.87 crore kg. In other words, a yearly deficit of 1,557,146 pigs or 9.20 crore kg of hog meat is recorded. Nevertheless, the market disruptions caused by ASF had a disastrous impact on small-scale pig producers in 2019. In order to mitigate the transmission of fatal diseases, officials eliminated certain animals after they contracted the disease, while others were slaughtered. Despite the substantial demand for pigs, the administration has maintained a farm in the Kelowna region ever since. A study was conducted to investigate the characteristics of pig manure in relation to its growth phases to gain a more comprehensive understanding of the nature of waste throughout a pig's lifetime (Table 2).

The total pig population experienced a moderate decline, as indicated by the 18th Livestock Census of India 2007. The Indian market for processed pork products heavily depends on imports due to its small size. Some local businesses manufacture processed meat products, including bacon and sausages; however, the industry is small and production is low. The Ministry of Food Processing Industries reports 3,600 slaughterhouses in India; however, most do not export (MoFPI). ASF, a critical pig illness, results in a substantial death rate and trade restrictions. ASF is more prevalent in Uganda, an impoverished nation with the highest pig population than in any other East African country.

3 Environmental Effects of Pig Farming and Some Technologies Adopted for the Treatment of Waste

At the Dutch Varkens Proefcentrum Sterksel, a pig farm that recycles waste was established. The pigs' diet is supplemented with algae cultivated using carbon dioxide and ammonia recovered from pig dung. Under the standards established by the American Samoa Environmental Protection Agency (AS-EPA) for piggery and waste use/disposal systems, all solid pig waste must be composted and flushed down with piggery water. The most prevalent configurations for suspended growth digesters that are used with animal waste are covered lagoons, mixed digesters (also referred to as continuously stirred tank reactors [CSTRs]), plug-flow digesters and anaerobic sequencing batch reactors (ASBRs) or upflow anaerobic sludge beds (UASBs). However, the primary method of swine manure management in the United States is land application as a crop fertiliser (López-Pacheco et al., 2019). For the majority of swine manures, plug-flow digesters are not suitable due to their high solid requirements (12–14% total solids), despite their widespread use with dairy waste.

The vast majority of organic nutrients, such as nitrogen, phosphorus and carbon molecules, are present in particles of extremely small size that are suspended in liquid swine. One of the most effective methods for removing nutritious elements from swine manure is to flocculate the mixture with polyacrylamide (PAM) and screen it, as per Paz Pérez-Sangrador et al. (2012). This distinguishes between liquid and solid components. The scarcity of arable land and the intensive pig raising have resulted in the water washing of pig manure off the floor and the facilitation of garbage collection. Consequently, collecting pig feces in solid manure form is challenging for small- and medium-scale farms. Pig farms necessitate an extensive quantity of water to ensure effective management. The wastewater that accumulates in vats contains a variety of contaminants, including pharmaceuticals such as antibiotics and antimicrobials, heavy metals like copper, iron and zinc, and environmentally harmful compounds like nitrogen, potassium and phosphorus. These contaminants include disease-causing bacteria like Corynebacterium and Salmonella (Koch et al., 2021).

The primary component of manure is fertilised animal waste, composed of approximately 60% feces and 40% urine (Vu et al., 2007). The three primary types of manure are solid, liquid and slurry. Solid manure is litter and feces that are collected from the floor, while liquid manure is a combination of urine, feces and water that are linked together during the scraping process. Slurry is a mixture of urine, feces and water. Trash is typically classified into three categories: The animal is housed in a system that collects its waste in liquid form, known as condensed manure (slurry). The floors are solid and sloped, regularly swept clean of waste. Some diluting is unavoidable due to the nature of wash water. The animals are housed on bedding material, and any liquid manure is collected separately. The solid dung is maintained in a separate stream. Solid dung is the sole excrement produced by animals confined to specific housing types. All other excrement, including that from the animals' bedding, is collected as farm yard manure (FYM).

Pig manure is typically disposed of in anaerobic lagoons in Western countries before being utilised as a fertiliser for crops (Szogi et al., 2015). The treatment of liquid livestock wastes, such as pig dung, is one potential application of aerobic thermophilic digestion. This procedure is purported to possess the following characteristics: simplicity, resilience, a higher reaction rate, nitrogen conservation and the potential for heat recovery, in addition to the ability to eliminate pathogens (Juteau, 2006). Although there are a few aerobic thermophilic plants for liquid manure, most are on a small scale and are only established for experimental purposes. These plants are found in the soils of the United States, Canada, the United Kingdom, Scandinavia, Czechoslovakia and Germany (Burton et al., 2003; Martens & Böhm, 2009). Anaerobic packed bed (APB) and UASB are the two most effective anaerobic digestion systems. A support is a defining characteristic of an APB reactor, with the primary purpose of enhancing the surface area of contact between the substrate and microorganisms. The aerobic biological process of composting entails the utilisation of microbes that are naturally present to convert biodegradable organic materials into a substance resembling humus (Jain et al., 2020). Composting simplifies the transportation and management of animal manure, and the nitrogen it contains is more stable and slow-release, enabling higher application rates (Sequi, 1996). The primary method of managing swine manure in the United States is the temporary storage and subsequent land application as a crop fertiliser. However, the recovery of energy and nutrients from manures before land application is becoming increasingly popular. In the past three decades, farmers in the United States have constructed approximately 100 full-scale animal waste digester systems. However, it is estimated that at least half of these digesters are currently inoperable (Amuah et al., 2022).

4 Assessment of Technologies Over Environmental Concerns of Piggery Wastewater

The treatment method or sequence of treatments (primary, secondary and tertiary treatments) for pig waste is determined by the method of waste collection, as illustrated in Table 3. The increasing demand for pork is making it more challenging to collect solid waste. Due to space constraints and intensive pig husbandry, waste collection has become more manageable, and soiled pigs are no longer an issue. Consequently, collecting pig feces as solid manure is a significant challenge for small- to medium-scale farms. Pig producers endeavour to accommodate the maximum number of swine in their overcrowded pens. The most prevalent approach to cleaning pig shelters is to squirt water to eliminate pig waste, feed remnants, bedding, urine and wash water.

The primary components of liquid manure are water and feces. Litter-free (dry) hog farms generate it as a waste product. Organic or phosphorus compounds are among the molecules that constitute liquid manure, which are present in the solid fraction. In addition to other components, the liquid fraction contains nitrogen compounds and mineral oxides of sodium, potassium and magnesium. The concentration of these compounds is contingent upon the manner in which they are raised and fed to hogs. Several proposed methods for treating piggery effluent are illustrated in Tables 4 and 5.

Table 3 Waste collection and treatment

Form of collection	Common treatment methods
Solid	Composting, fish feed
Liquid	Anaerobic ponds, anaerobic filter, covered lagoon
Slurry	Anaerobic digestion

Table 4 Treatment efficiency of each technology

Treatment technology			Removal	Efficiency (%)			
Medium-scale farm	BOD	COD	Suspended solids (SS)	Nitrogen	TKN	Phosphorus	*Copper*
Stabilisation pond	78.52	68.24	78.02	49.25	49.45	50.93	75.28
Anaerobic filter (AF)	82.34	77.50	81.55	61.96	64.35	62.71	90
Lagoon	69.83	75.98	84.48	53.70	48.68	50.12	82.80
Fixed dome biogas digester	75.86	65.53	77.26	38.56	36.97	43.58	85.56
UASB + channel digester	85.47	70.74	82.86	46.72	49.41	30.55	76.96

Note UASB: upflow anaerobic sludge bed; COD: chemical oxygen demand; BOD: biochemical oxygen demand; TKN: total Kjeldahl nitrogen
Source Asian Development Bank (2024)

Table 5 Piggery wastewater treatment technology

	Anaerobic pond	Facultative pond	Aerobic pond
Depth	Fairly deep	Deepest	Shallowest of ponds
Stage of treatment	Primary	Effluent from anaerobic pond sent here for further BOD removal	Last step in the series of pond treatment
Treatment process	Solids and settleable organics settle to the bottom forming sludge, which is digested by anaerobic microorganisms	Top layer aerobic, bottom layer anaerobic: algae growing on the surface provide the water with oxygen	Retention of stabilised solids and inactivation of pathogenic microorganisms via heating, rise of pH and solar disinfection
BOD removal	50–85%	80–95%	60–80%

Source Asian Development Bank (2024)

5 Positive Perspectives for Applicability of Anaerobic Piggery Wastewater Treatment

Sugar beet, slaughterhouse, starch, brewery and other industrial wastewaters are used extensively for anaerobic treatment. The temperature of pig waste is a major factor determining whether anaerobic treatment is suitable for piggery wastewater. At 35 °C, the functioning of mesophilic anaerobic bacteria is at its peak. At lower temperatures, bacterial activity decreases, resulting in lower treatment performances (Show et al., 2010).

As illustrated in Fig. 2, most emerging nations experience tropical or subtropical temperatures, which necessitates the primary focus of these nations on anaerobic piggery wastewater treatment. Bacteria are cultured in fixed-film systems by adhering to various supporting media, including sponge cubes, pall rings, plastic pipes and other structures (Setiyawan et al., 2023). Fixed film anaerobic systems are frequently referred to as ‘anaerobic filters’ due to the necessity of ‘filtering’ the waste through the reactor’s filter media. The filter media is designed with a specific surface area characterised by extensive holes to prevent clogging or plugging caused by manure solids (Gullicks et al., 2011). Anaerobic filters are effective for flush manure management systems and other low-solid manures. The system may digest more manure in less time than conventional digesting methods due to the bacteria’s ability to remain stationary, which enables the use of a smaller reactor (El Mashad et al., 2023). In order to expedite the digestion process, digestion systems are either heated to a specific temperature or operate at room temperature. The digester system is either heated or a portion of the operation’s energy needs is satisfied by converting the biogas that is generated (Deng et al., 2023).

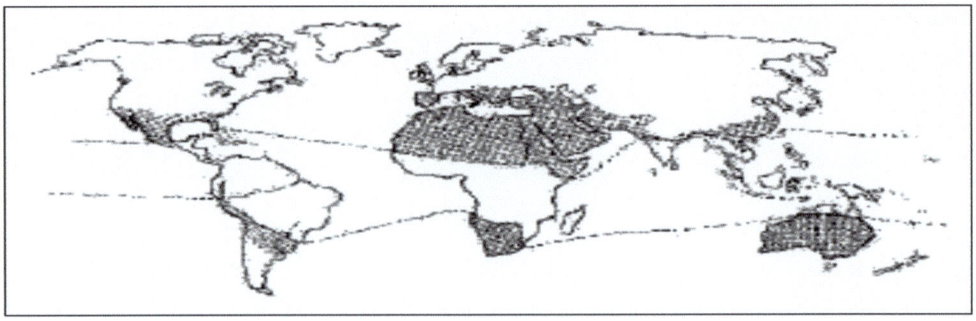

Fig. 2 World temperature zones (Van Haandel & Lettinga, 1994)

6 Conclusion

The global population is expanding at a rate that is comparable to that of the swine waste sector. The proliferation of concentrated pig farms, which have herds of thousands, is a significant development. These augmented herds are accompanied by substantial volumes of waste, which includes organic matter, inorganic nutrients and gaseous emissions. Major infections, including leptospirosis, gastroenteritis, cholera, dysentery, infectious hepatitis and salmonellosis, are prevented from spreading to humans by managing pig manure in an environmentally appropriate manner. Humans are susceptible to infections caused by pathogens found in pig excrement through the skin, eyes, mouth and nose. These infections are frequently detected in bodies of water, such as streams, stormwater and other bodies of water, as a result of the improper disposal of pig manure.

The literature suggests that the treatment of piggery wastewater has been the subject of numerous studies. The progressive accumulation of non-biodegradable particles, which degrades the bed structure, is the primary cause of the physical constraints of anaerobic filters. Anaerobic filters are not suitable for wastewaters that contain numerous particles, as this ultimately leads to flow obstruction and channelling. Anaerobic degradation is also significantly influenced by numerous extrinsic factors. The consumption of pig meat is on the rise, particularly in the northeastern states of India, which has resulted in an unsettling situation with respect to the safe disposal of effluents from farms. The management of pig manure in India fails to consider the negative environmental and human health consequences that are linked to the untreated disposal of wastewater.

References

Amuah, E. E. Y., Fei-Baffoe, B., Sackey, L. N. A., Douti, N. B., & Kazapoe, R. W. (2022). A review of the principles of composting: Understanding the processes, methods, merits, and demerits. *Organic Agriculture, 12*(4), 547–562. https://doi.org/10.1007/s13165-022-00408-z

ASAE. (2005). *Manure production and characteristics*. ASABE. https://elibrary.asabe.org/abstract.asp?aid=32018&t=2

Bank, A. D. (2024). *Pollution characterization and quantification in the agriculture sectors* (Issue April).

Burton, C. H., Turner, C., Beck, J. A. F., Martinez, J., Martens, W., Pahl, O., Piccinini, S., & Svoboda, I. (2003). *Manure management: Treatment strategies for sustainable agriculture* (2nd edn.). Silsoe Research Institute.

Deng, L., Zheng, D., Zhang, J., Yang, H., Wang, L., Wang, W., He, T., & Zhang, Y. (2023). Treatment and utilization of swine wastewater—A review on technologies in full-scale application. *Science of The Total Environment, 880*, 163223. https://doi.org/10.1016/j.scitotenv.2023.163223

Department of Aminal Husbandry and Dairying, G. of I. (2019). *20 TM LIVESTOCK CENSUS-2019.*

Dohányos, M., Zábranská, J. and Jeniícek, P. (1997). Innovative technology for the improvement of the anaerobic methane fermentation. *Water Science and Technology, 36*(6), 333–340. https://doi.org/10.1016/S0273-1223(97)00540-4

El Mashad, H. M., Barzee, T. J., Franco, R. B., Zhang, R., Kaffka, S., & Mitloehner, F. (2023). Anaerobic digestion and alternative manure management technologies for methane emissions mitigation on Californian dairies. *Atmosphere, 14*(1). https://doi.org/10.3390/atmos14010120

FAO. (2019). The state of food and agriculture 2019. *FAO*. https://doi.org/10.4060/CA6030EN

FAO. (2021). World food and agriculture—statistical pocketbook 2021. *FAO*. https://doi.org/10.4060/cb4478en

Gullicks, H., Hasan, H., Das, D., Moretti, C., & Hung, Y.-T. (2011). Biofilm fixed film systems. *Water,* 3(3), 843–868. https://doi.org/10.3390/w3030843

Jain, M. S., Paul, S., & Kalamdhad, A. S. (2020). Kinetics and physics during composting of various organic wastes: Statistical approach to interpret compost application feasibility. *Journal of Cleaner Production, 255*, 120324. https://doi.org/10.1016/j.jclepro.2020.120324

Juteau, P. (2006). Review of the use of aerobic thermophilic bioprocesses for the treatment of swine waste. *Livestock Science, 102*(3), 187–196. https://doi.org/10.1016/j.livsci.2006.03.016

Koch, N., Islam, N. F., Sonowal, S., Prasad, R., & Sarma, H. (2021). Environmental antibiotics and resistance genes as emerging contaminants: Methods of detection and bioremediation. *Current Research in Microbial Sciences, 2*, 100027. https://doi.org/10.1016/j.crmicr.2021.100027

López-Pacheco, I. Y., Silva-Núñez, A., Salinas-Salazar, C., Arévalo-Gallegos, A., Lizarazo-Holguin, L. A., Barceló, D., Iqbal, H. M. N., & Parra-Saldívar, R. (2019). Anthropogenic contaminants of high concern: Existence in water resources and their adverse effects. *The Science of the Total Environment, 690*, 1068–1088. https://doi.org/10.1016/j.scitotenv.2019.07.052

Martens, W., & Böhm, R. (2009). Overview of the ability of different treatment methods for liquid and solid manure to inactivate pathogens. *Bioresource Technology, 100*(22), 5374–5378. https://doi.org/10.1016/j.biortech.2009.01.014

Paz Pérez-Sangrador, M., Cristina León-Cófreces, M., Acítores-Benavente, M., & Cruz García-González, M. (2012). Solids and nutrient removal from flushed swine manure using polyacrylamides. *Journal of Environmental Management, 93*(1), 67–70. https://doi.org/10.1016/j.jenvman.2011.07.020

Sanchez, P. A., & Leakey, R. R. B. (1997). Land use transformation in Africa: Three determinants for balancing food security with natural resource utilization. In M. K. van Ittersum & C. S. van de Geijn (Eds.), *Perspectives for agronomy* (Vol. 25, pp. 19–27). Elsevier. https://doi.org/10.1016/S0378-519X(97)80004-9

Sequi, P. (1996). The role of composting in sustainable agriculture. In M. de Bertoldi, P. Sequi, B. Lemmes, & T. Papi (Eds.), *The science of composting* (pp. 23–29). Springer Netherlands. https://doi.org/10.1007/978-94-009-1569-5_3

Setiyawan, A. S., Nur, A., Fauzi, M., Oginawati, K., & Soewondo, P. (2023). Effects of different polymeric materials on the bacterial attachment and biofilm formation in anoxic fixed-bed biofilm reactors. *Water, Air, & Soil Pollution, 234*(3), 147. https://doi.org/10.1007/s11270-023-06174-2

Show, K. Y., Tay, J. H., & Hung, Y.-T. (2010). *Global perspective of anaerobic treatment of industrial wastewater.* In L. K. Wang, J.-H. Tay, S. T. L. Tay, & Y.-T. Hung (Eds.), *Environmental bioengineering* (Vol. 11, pp. 773–807). Humana Press. https://doi.org/10.1007/978-1-60327-031-1_23

Szogi, A. A., Vanotti, M. B., & Ro, K. S. (2015). *Methods for treatment of animal manures to reduce nutrient pollution prior to soil application* (pp. 47–56). https://doi.org/10.1007/s40726-015-0005-1

Turner, C., & Burton, C. H. (1997). The inactivation of viruses in pig slurries: A review. *Bioresource Technology, 61*(1), 9–20. https://doi.org/10.1016/S0960-8524(97)84693-7

Van Haandel, A., & Lettinga, G. (1994). *Anaerobic sewage treatment: A practical guide for regions with a hot climate.* Wiley.

Vu, T. K. V, Tran, M. T., & Dang, T. T. S. (2007). A survey of manure management on pig farms in Northern Vietnam. *Livestock Science, 112*(3), 288–297. https://doi.org/10.1016/j.livsci.2007.09.008

A Compendium of Existing Sewage Treatment Techniques at India's Premier Higher Technical Educational Institutes

Sugato Panda, Mayur Shirish Jain, Sai Krishna Akash Ramineni, and Polisetti Sai Meghan

1 Introduction

Higher education institutions (HEIs) have significant freedom for managing natural resources; most HEIs have residential campuses equipped with essential amenities that often make them mini cities (Mahatma Gandhi National Council for Rural Education [MGNCRE], 2019). Most of these institutes are located at remote locations from the main towns, necessitating more assistance from Urban Local Bodies (ULBs). Hence, these institutes must have efficient technologies to maintain adequate campus sanitation standards. In the Indian context, 67% of sewage treatment plants (STPs) in urban areas are operational, 6.25% are non-operational, 16.74% are under construction, and 9.93% are proposed for construction (Central Pollution Control Board [CPCB], 2021). Similarly, just 33.33% of sewage is treated by Indian STPs (Fig. 1a), which calls for stepping up treatment processes. As all HEIs are mandated to follow the sustainable development goals (SDG), clean water and sanitation are of utmost priority while providing students with a viable and plush environment. The early reconciliation with a lower standard of living may have an everlasting effect on students and researchers, which becomes apparent in their professional performance and contributions to national interests. Therefore, the emphasis should be on maintaining the quality of the physical environment.

Sewage treatment processes are classified into physical, chemical and biological, as mentioned by Bolong et al. (2022), with maximum treatment technologies in India being sequencing batch reactors (SBRs) and activated sludge processes (ASPs) (Fig. 1b). The surge in urban population is likely to trigger a spontaneous increase in the production of municipal wastewater. As India expands its wastewater treatment facilities to accommodate this growth, it may lead to the generation of human-made greenhouse gas emissions and an upsurge in energy usage (Singh & Kansal, 2018). Compared to India, the United States and China have higher water treatment capacities, with the United States having greater scope for water reuse (Fig. 2a).

Even though the United States, China, and European countries have several elite institutions, they still need to publish evidence of wastewater treatment technology practiced on campus premises (Fig. 2b). As Negi and Chandel (2021) argue, STPs consume approximately 3% of the total electricity used in the United States for their operation and maintenance, a study about wastewater and energy nexus would be a steady overlap towards SDG 6 and 7. A study was conducted on reclaimed water quality from a 480 m^3/day STP at Tunghai University, Taiwan, to understand alternative disinfection practices to promote the utilisation of pathogen-free water on the campus (Kuo et al., 2023). Another STP for research is operational at the University of Stuttgart (Teaching and Research STP [LFKW]., n.d.) near the village of Büsnau with a treatment capacity of 2,000 m^3/day for 10,000 inhabitants. Moreover, one of the distinctive features of LFKW is that it is currently the sole treatment plant owned by a university, at least in Europe. In the United States, both Pennsylvania State University (2023) and the University of California, Davis own on-campus wastewater treatment facilities. The water reclamation facility (WRF) at Pennsylvania State University (2023) was designed in 1983 and has been treating around 3 million gallons per day and utilises ASP.

In the case of the University of California, Davis, the southern campus at Trotter Lane is mainly managed by institute's professors and students. The plant started to operate in 2000 with an initial setup cost of US$15.3 million (LocalWiki, n.d.). The treatment plant uses ultraviolet (UV) light in its disinfection process. Additionally, it's worth mentioning that the university has constructed an anaerobic phased solids digester, which converts solid organic waste into electricity

S. Panda (✉) · M. S. Jain · S. K. A. Ramineni · P. S. Meghan
Department of Civil Engineering, Indian Institute of Technology Indore, Indore, Madhya Pradesh, India
e-mail: phd2201104004@iiti.ac.in

© Capital Publishing Company, New Delhi, India 2025
I. Haq et al. (eds.), *Water and Wastewater*, https://doi.org/10.1007/978-3-031-94253-2_10

Fig. 1 **a** Skewness between generation, operation, and maintenance capacities of Indian STPs. **b** Prevalent STP technologies in the Indian context (CPCB, 2021)

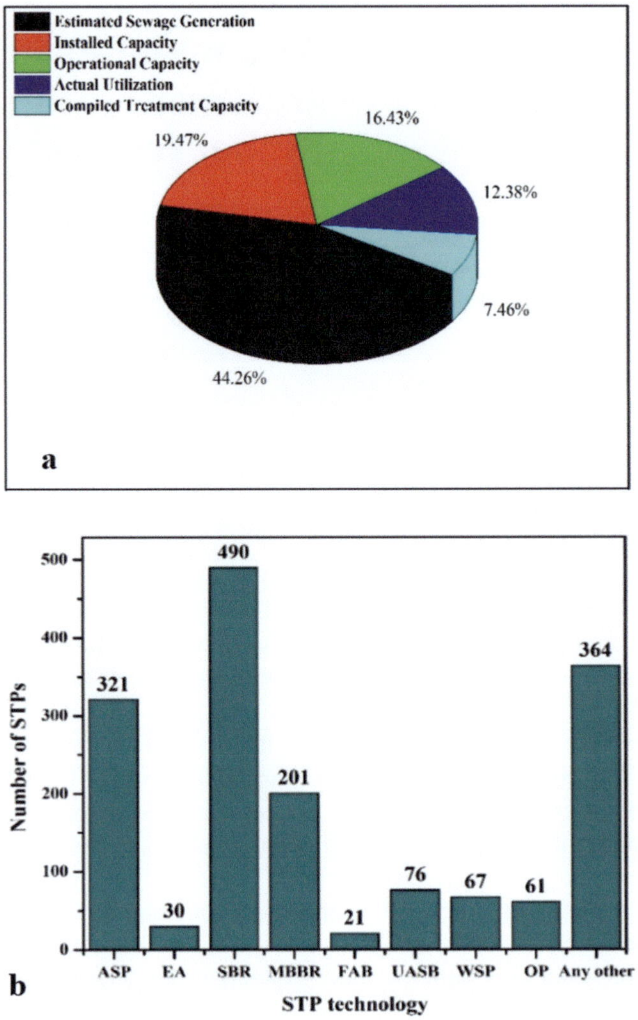

and fibre products. The facility can treat a maximum of 2.8 million gallons of waste daily.

Another example of a decentralised STP is the University of Queensland's SBR-based STP in Australia which treats sewage from the regional campus, research institutes, and restaurants. The design incorporates chemical dosing systems to handle pH correction, control precipitating phosphorus, and conditioning the sludge against dewatering (Aerofloat, 2022). Furthermore, the plant encounters 2 mg/L of 5-day biochemical oxygen demand (BOD_5) and 8 mg/L of suspended solids in the outlet, with trace amounts of nitrates and phosphorus, demonstrating the plant's effectiveness.

To date, India has 55,000 HEIs, of which 43,796 are colleges (either affiliated or recognised by universities; Statista, 2021); among these, there are 23 Indian Institutes of Technology (IITs), 31 National Institutes of Technology (NITs) and one Indian Institute of Science (IISc), which come under the domain of premier institutions. These premier institutions were set up to provide cutting-edge technical and economical solutions to propel the country's research output. Although most scientists and academicians from these institutes contribute voraciously towards environmental research, not all the institutes mention details and effectiveness of their sewage treatment technologies on their campuses. At the outset of SDG 6, where the government is pushing the idea of decentralised sewage treatment, the success stories from in-house plant operations motivate other institutions to contribute to sharing the load of municipalities.

Hence, the present study aimed to collate all the existing techno-managerial details of in-house treatment facilities of these institutes available in public domain. Further, it compares these technologies individually regarding efficiency and operational challenges and suggests a standard operating procedure for the steady functioning of decentralised STPs.

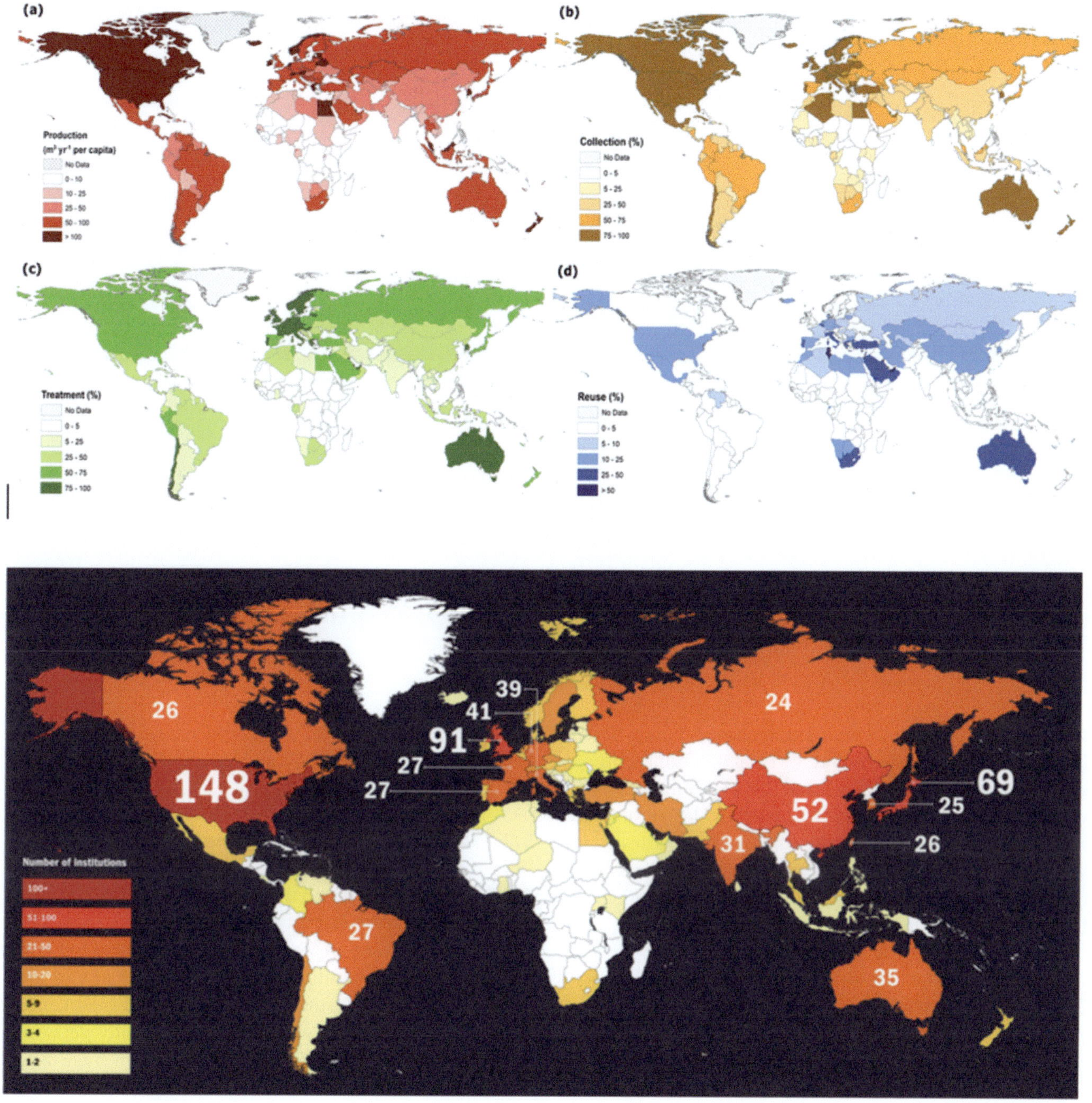

Fig. 2 **a** Worldwide production, collection, treatment and reuse (Jones et al., 2021). **b** Countries ranked as per the 'Times Higher Education' (World Economic Forum [WEF], 2016)

2 Methodology

The National Institutes Ranking Framework (NIRF) was used for the demarcation of institutes based on the quality of education and facilities. Further, the institutes were pooled into three categories: IITs, NITs, and private institutions. A schematic flow diagram represents the methodology in greater detail (Fig. 3).

Primarily, institute websites, newspaper reports, and project reports were referred for information. Among the IITs, IIT Kharagpur, IIT Bombay, IIT Madras, IIT Roorkee (n.d.), IIT Mandi (n.d.), and IIT Gandhinagar have details on their websites or review reports. Details of IIT Indore were collected through personal visits and interactions. In the case of NITs, NIT Rourkela has a documented mention of having an STP on its campus, with most of the NITs

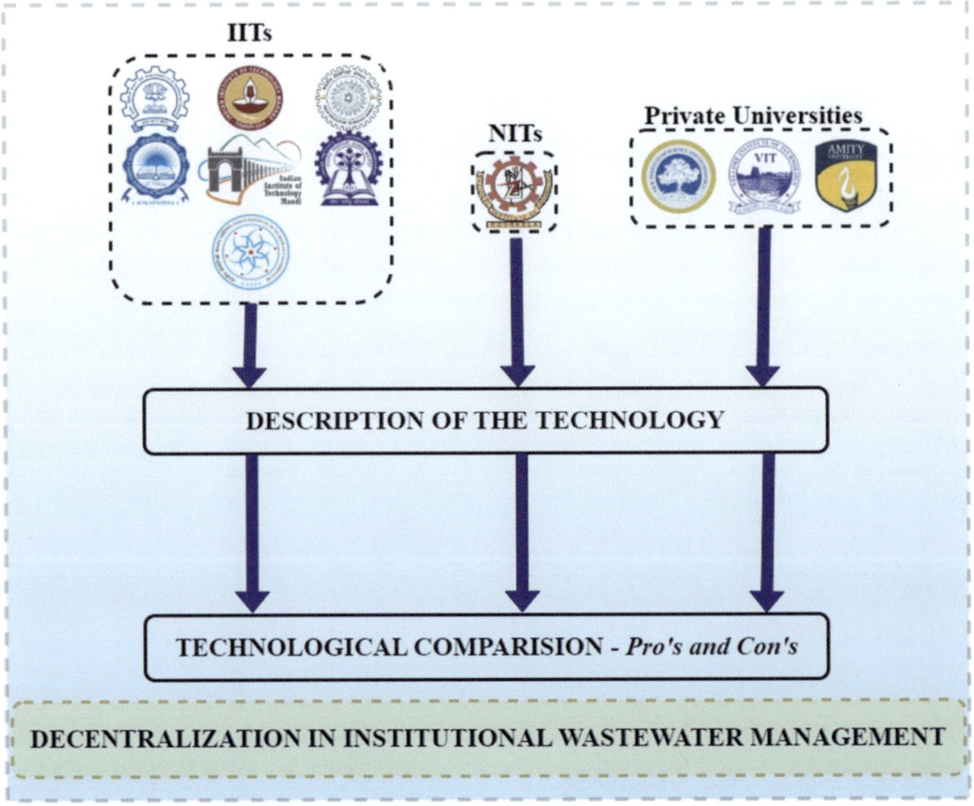

Fig. 3 Flow diagram of methodology for the study

still in the planning and construction phase. Finally, Vellore Institute of Technology Vellore (VIT, n.d.), Amity University, Gurugram, Haryana (n.d.), and SRM University Chennai were the only three institutions that provided information regarding technologies utilised by their STPs. The discussion was mainly based on the overview and details of plants, aided by design philosophies and techno-economics to understand the long-term viability of decentralisation.

3 Indian Institutes of Technology

3.1 Indian Institute of Technology Bombay

In 2013, the Indian Institute of Technology Bombay (IIT Bombay), in collaboration with the Centre for Science and Environment (CSE, n.d.), established a constructed wetland (CW)-aided STP on the campus. The overall treatment arrangement is for a capacity of 25 kilolitres per day (KLD), comprising an initial settling tank, succeeded by a 400 m^2 planted filter bed. This bed is covered with wetland vegetation and is structured as a horizontal sub-surface flow (HSSF) mechanism (CSE, n.d.). Further, the dimensions of the wetland were 13 × 3 × 0.6 m, with a gravel bed layer and a more extensive gravel padding of 50–100 mm in diameter at both inlet and outlet (Chatterjee, 2016). The water then reaches a water regulation chamber for disinfection before final collection. The treated wastewater from the system is then discharged into Powai Lake.

3.2 Indian Institute of Technology Madras

3.2.1 Need for Decentralised Wastewater Treatment Facility

The Indian Institute of Technology Madras was set up in 1959 and became an institute of national importance in 1961. Spread over 248 acres, the campus has a total water demand of 3.5 megalitres per day (MLD) and is mainly dependent on the Chennai Metro Water Supply and Sewerage Board (CMWSSB) for its water requirements. However, because the water supply from the CMWSSB was significantly depleted during the summer season, the need appeared to set up an independent wastewater and water treatment source.

3.2.2 Description of Technology

In order to augment water supply, a 4-MLD complete wastewater treatment facility and a rainwater harvesting system were established on 26 January 2017. The facility consists of primary, secondary, and tertiary treatment services. Before the biological treatment, the wastewater passes through the screening tank and reaches the grit chamber, which precedes an equalisation tank. It is then followed by an SBR arrangement for biological treatment,

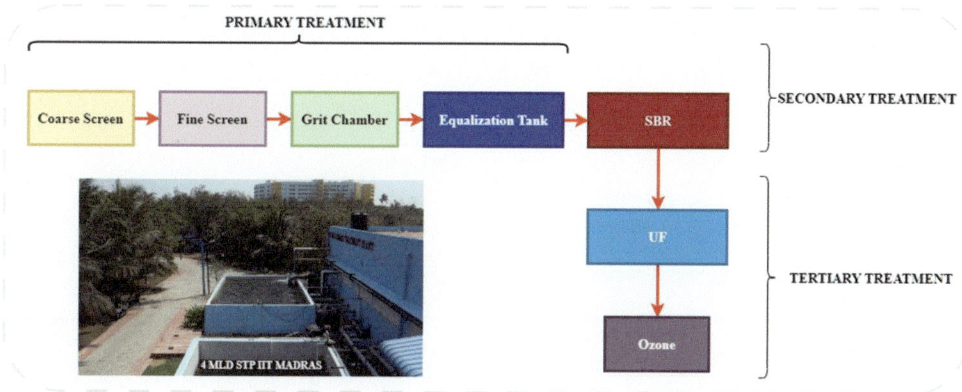

Fig. 4 4-MLD sewage treatment plants at IIT Madras (Krishnamurthy, 2019)

followed by an ultrafiltration (UF) system as a tertiary treatment. Prior to leaving the plant, the water is passed through an ozonation facility for disinfection and finally utilised in toilet flushes, horticulture, and cooling towers for a centralised AC system. The residual water is used for recharge of well after passing through infiltration galleries. A schematic as discussed is mentioned in Fig. 4.

3.3 Indian Institute of Technology Roorkee

Established in 1847 by James Thomason, the transition from the College of Civil Engineering, Roorkee, to the Indian Institute of Technology Roorkee (IIT-R) throughout these 175 years has been inspiring. Despite contributing to various engineering domains, the college has been the beacon to establish a decentralised STP on the campus. According to IIT-R's institute works website, the plant has been operational since 2019 at Solani Kunj (IIT Roorkee, n.d.). At present, the 3-MLD plant, designed according to the SBR technology, has been operating at 66% of its design capacity because of ongoing work establishing new sewerage networks (Fig. 5a).

3.3.1 Description of Technology

The SBR technology employs a fill-and-draw cyclical protocol, including aeration and non-aeration sequencing, to obtain biological treatment and sedimentation/clarification in the same basin. The system provides high-quality effluent for safe discharge or non-potable reuse with minimum space and power requirements. Additionally, it has access to odour control and a tertiary treatment facility because of settling down in a residential area. The deodorisation facility connects with a sump well as well as a pre-treatment facility. The STP is crafted meticulously with a focus on aesthetics, and all the processing units are enclosed within closed structures. Cutting-edge tertiary treatment methods, including fibre disc filtration and UV disinfection, are implemented to ensure safety of the treated water for reuse for campus horticulture. At the same time, sludge is repurposed as an organic fertiliser. Overall, the STP has two phases: (a) fill and aeration phase (90 min) and (b) settling (45 min) and decanting phase (45 min). Regarding performance, the plant has demonstrated great flexibility in handling influent during monsoons, with excellent effluent quality at outlet (Fig. 5b and c).

3.4 Indian Institute of Technology Indore

Established in the cleanest city of India, IIT Indore was set up in 2009. Starting in February 2016, IIT Indore has been operational at its permanent campus, housing various administrative offices, material management section, finance and accounts section, School of Basic Sciences, School of Humanities and Social Sciences, and School of Engineering as well as both basic science laboratories and engineering laboratories. Besides having a water treatment facility, the institute has a decentralised 650-KLD MBBR- aided STP housed on the campus. (Fig. 6).

3.4.1 Description of Technology

Wastewater from hostels and residential areas is collected in an equalisation tank; it then passes into an aeration tank and is subjected to microbial degradation. The MBBR functions as a continuous-flow aerobic system, housing a population of activated microorganisms that effectively attaches into a filter media and stabilises organic substances. This tank combines wastewater with a thriving community of microorganisms, primarily encompassing bacteria and protozoa. These microorganisms degrade organic materials present in the wastewater through aerobic processes, forming carbon dioxide, water, fresh cells, and various other by-products (Bhargava, 2016). Post-clarification, the water passes through an activated carbon filter, and the sludge is dried in the open and used as compost. The treated water serves dual purposes: it is employed for horticultural activities and then cycled back into storage tanks in hostels. These

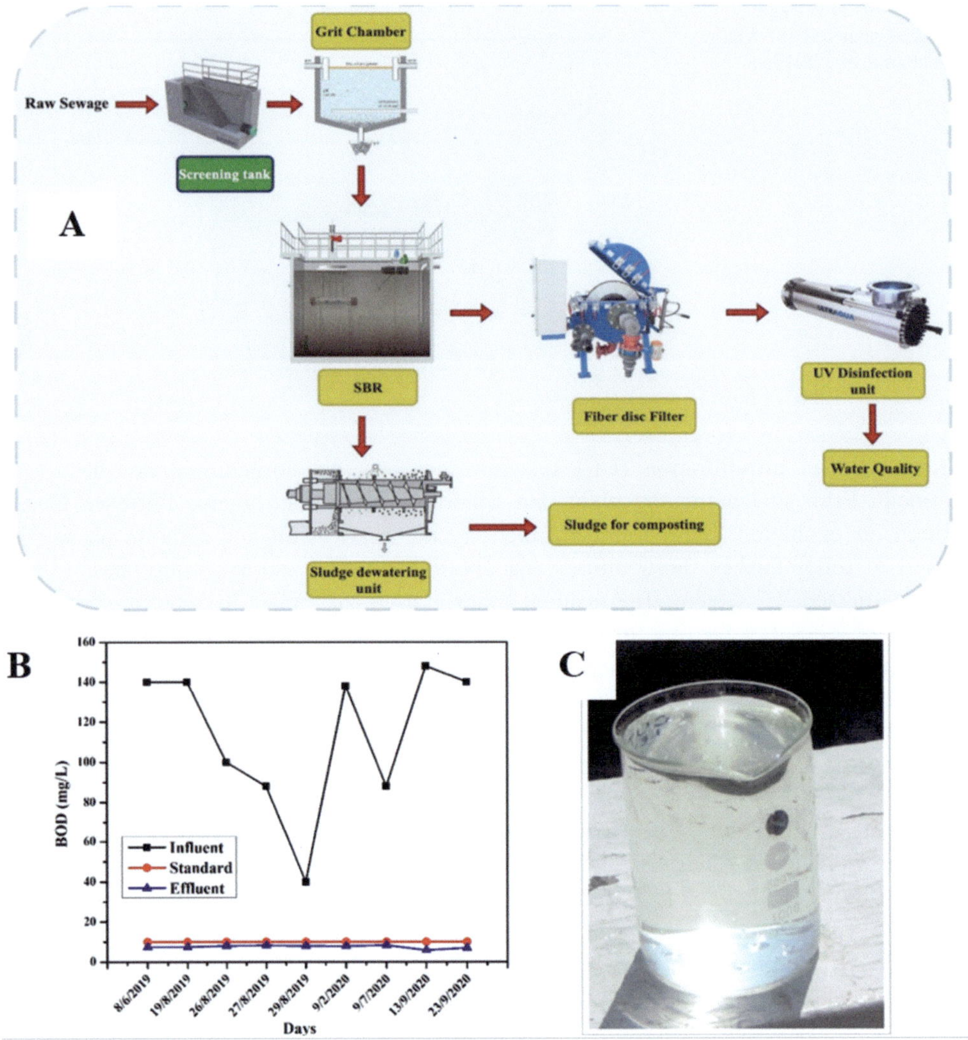

Fig. 5 **a** Schematic of sewage treatment process at IIT Roorkee. **b** BOD removal efficiency. **c** Sparkling effluent (IIT Roorkee, n.d.)

tanks serve as reservoirs for reused flush water. The institute has an indoor laboratory beside the treatment plant for testing of all basic physico-chemical parameters. For biological parameters, samples are sent to a third-party partner laboratory.

3.5 Indian Institute of Technology Mandi

IIT Mandi, situated in Kamand Valley of Himachal Pradesh, is one of the eight new IITs established in 2009. It has five STPs with various technologies in its north and south campuses (IIT Mandi, n.d.). The capacities and technology details are mentioned in Table 1. Similar to the above-described IITs, the prevalent technology on most of its campuses is SBR.

3.6 Indian Institute of Technology Gandhinagar

3.6.1 Description of Technology

Sewage treatment plant at IIT Gandhinagar, established in 2008, is a five-stage process. Wastewater is collected through the force of gravity using unplasticised polyvinyl chloride (PVC-U) SN-8 pipes without plasticising or pressure assistance, leading to sewage pumping stations. The transportation to 6-MLD STP occurs via k7 ductile iron pipes as illustrated in Fig. 7. The raw sewage passes through a tube settler, where all the suspended solids are separated by gravity. Post-settling, the sewage passes onto an anaerobic baffled reactor (ABR), where suspended and dissolved solids are degraded (Parikh et al., 2019). Following this, the sewage passes through a planted gravel filter bed having *Canna indica* (herbaceous shrub) for root zone treatment. Residual pollutants from ABR flow through the plant's root zone and are adsorbed. The

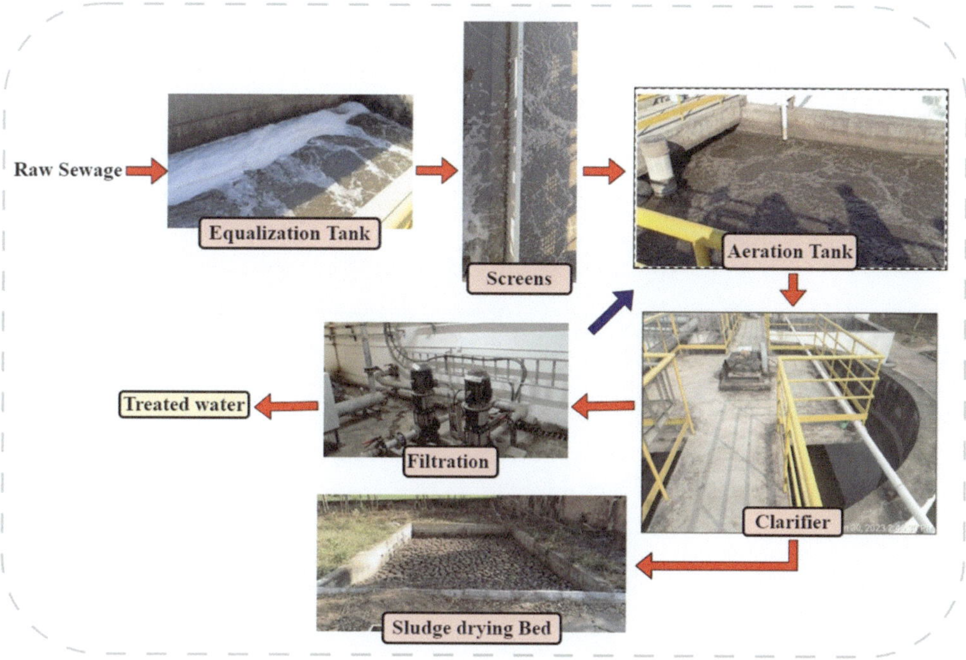

Fig. 6 Schematic representation of 650-KLD STP on IIT Indore campus

Table 1 Details of scale and technology of sewage treatment plants at IIT Mandi, Himachal Pradesh, India

S. no	Capacity	Campus	Technology
1	450 KLD	North	SBR
2	75 KLD	South	MBBR
3	75 KLD	South	MBR
4	25 KLD	Gharpa South	SBR
5	125 KLD	South	SBR

Notes SBR: sequencing batch reactor; MBR: membrane bioreactor; MBBR: moving bed biofilm reactor

fourth and fifth stages are filtration systems (ultrafiltration and superfiltration) that further reduce suspended solids and turbidity. Maximum BOD reduction occurs at this stage. Before discharging the treated wastewater, it passes through a UV-based disinfection system. Although based on the report by Parikh et al. (2019), the campus does not utilise UF systems, and the water is pumped directly to storage tanks for campus irrigation and flushing.

3.7 Indian Institute of Technology Kharagpur

The Indian Institute of Technology Kharagpur (IIT Kharagpur) is one of the most prestigious institutions in India and is the first among all IITs established in 1951. Unlike other IITs, IIT Kharagpur has multiple surrounding rural districts, such as West Midnapore, Bankura, and Purulia, ranked poorly according to the human development index (Das & Bokshi, 2017). An article published in *Indian Express* in 2018 described that IIT Kharagpur was planning to establish a sewage treatment facility to treat 1.35 MLD of sewage into 1.2-MLD of potable water. Apart from this, no other information on STPs is currently available from the institute.

4 National Institutes of Technology

4.1 National Institute of Technology Rourkela

NIT Rourkela in Odisha was established in 1961 and is among the 31 NITs established in the country. Although the institute does not officially mention the existence of an STP on its campus, Abhinav (2015) reports the presence of a 0.2-MLD STP near the Vikram Sarabhai Hall of Residence for the treatment of wastewater from hostels and residential areas. Considering the incessant generation of sewage sludge, no hardcore literature discusses the exact technology practiced there.

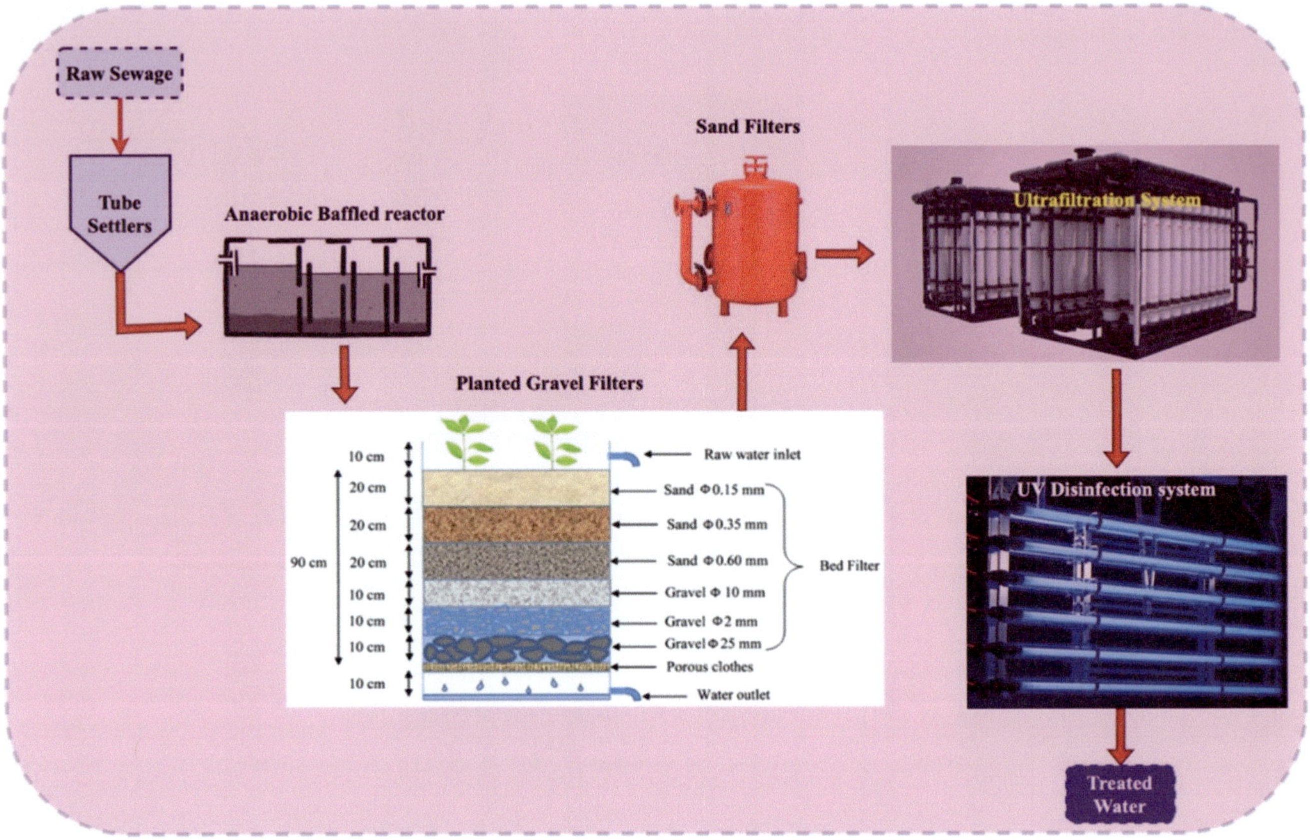

Fig. 7 Process flow diagram of a 6-MLD STP on IIT Gandhinagar campus

5 Private Universities

5.1 SRM University Chennai

The SRM University Chennai was set up in 2002 in the Kattankulathur district of Tamil Nadu. To cater to the wastewater treatment needs, the institute has three plants with capacities of 1.5-MLD, 2.6-MLD, and 3.7-MLD (SRM Institute of Science and Technology [SRMIST], 2021). Primarily, the biological treatment is done using the moving bed biofilm reactor (MBBR) technology. After screening, wastewater passes through MBBR aeration unit. Sewage from MBBR process is collected by a clarifier, and the overflow goes through pressure sand filter (PSF) and activated carbon filter (ACF). The treated water is passed for horticulture, and a part of the flow after ultrafiltration is used for toilets. The overall removal efficiency of COD is 98%. A schematic illustration of the process is provided in Fig. 8.

5.2 Amity University Haryana

The Gurugram (Haryana) campus of Amity University (n.d.) has two STPs and two different effluent treatment plants (ETPs) with a design capacity of 9 MLD and 0.05 MLD (Fig. 9). As discussed previously, the treated water is applied for flushing, gardening, and maintenance of cooling unit and administrative buildings.

5.3 Vellore Institute of Technology

The Vellore Institute of Technology, established in 1984, is one of the best private engineering institutes in the country. The institution is equipped technologically, has access to a 4.5-MLD STP, and claims 90% wastewater recycling (Vellore Institute of Technology [VIT] University, n.d.). This recycled water is used for horticulture, flushing, washing of vehicles and as an input to chiller plant. However, the type of technology used for plant setup on the campus is not mentioned.

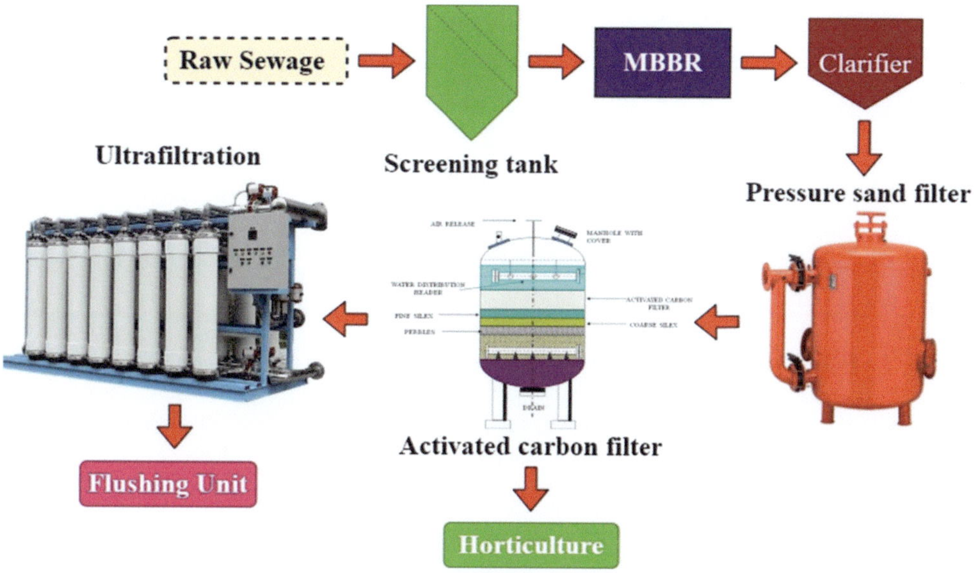

Fig. 8 Process schematics of sewage treatment plant on SRM Chennai campus

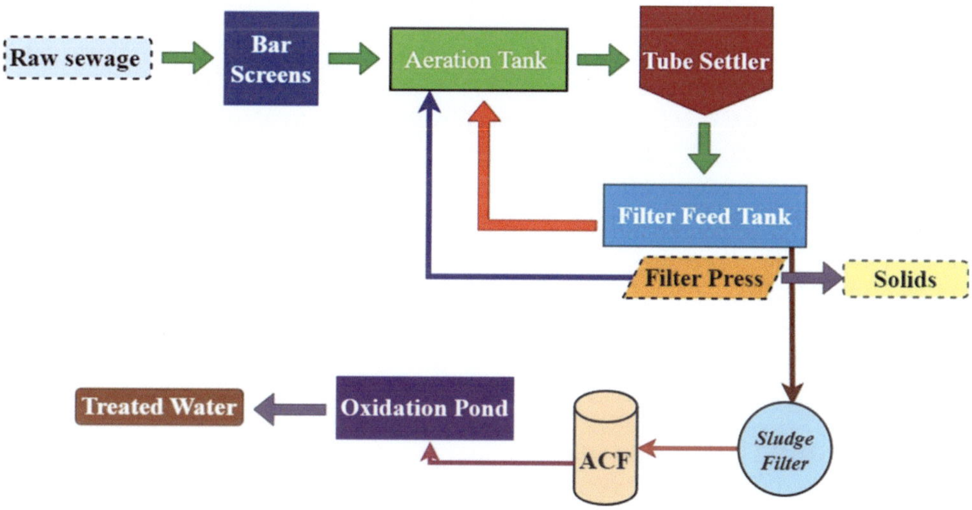

Fig. 9 Process schematics of STP at Amity University (n.d.)

6 Techno-Economic Discussion on Prevalent Technologies

As discussed previously, it is evident that the dominant technologies are SBR, MBBR, ASP, and CW. Therefore, detailed techno-economic pros and cons are necessary to understand the long-term viability of decentralised adaptability in the Indian landscape. Although minimal details are available on the setup costs of these plants, it has been attempted to extrapolate generalised information on the capital and operational expenditures for setting up of these technologies. A thorough analysis is necessary not only for systematic financial decision-making but also for resource optimisation, such as energy consumption, chemical usage, and sludge disposal.

6.1 Constructed Wetlands

The cost-effectiveness of constructed wetlands (CWs), compared to traditional systems, and their lower energy demands than conventional methods highlight their significance as sustainable sanitation alternatives. This is particularly notable when contemplating CWs as decentralised options for dispersed communities. Numerous research studies have underscored the economic benefits of CWs, including reduced construction and operational expenses, compared to conventional wastewater treatment plants (Masoud et al., 2022). CWs are low-cost processes that treat wastewater with low operation and maintenance costs, with operation costs corresponding to 1–2% of construction costs (Parde et al., 2021). Compared to the conventional ASP, the

Table 2 Techno-economic comparison of constructed wetlands to available literature

s. no	Constructed wetland (country)	Cost (INR equiv.) (CAPEX/OPEX dynamics)	References
1	IIT Bombay, India	OPEX = 12% of CAPEX	CSE (n.d.)
2	Two-stage wetland system in Austria	OPEX = 3.8% of CAPEX	Langergraber et al. (2018)
3	CW in Moldova, France	OPEX = 2.5% of CAPEX	Masi et al. (2017)
4	CW in Gorgona, Italy	OPEX = 0.5% of CAPEX	Cross et al. (2021)
5	CW in Jesi, Italy	OPEX = 0.5% of CAPEX	Masoud et al. (2022)

Notes CAPEX: capital expenditures; OPEX: operating expenditures; CW: constructed wetland

Table 3 Probable CAPEX estimates of MBBR plants installed at IITs considering linearity of scales with increasing capacity

S. no	Sewage treatment plant (MBBR)	CAPEX (INR)
1	SRM University, Chennai (1.6-MLD)	2,56,00,000
2	SRM University, Chennai (2.5-MLD)	4,00,00,000
3	SRM University, Chennai (3.7-MLD)	5,92,00,000
4	IIT Mandi, Himachal Pradesh (75 KLD)	12,00,000

construction costs are 30% lesser, and the operational costs are similar to the construction costs (Eriks, 2011). A comparative differentiation of CW at IIT Bombay is made with available literature (Table 2). Although the relatively higher operating expenditures (OPEX) at IIT Bombay may correspond to the cost incurred for research and development, as the report was published 10 years back, it is expected that the process is much more economical currently and hence implemented in different areas of Mumbai municipality (Bombay Municipality Corporation or BMC).

6.2 Sequencing Batch Reactor

The advantage of utilising SBR technology is mainly attributed to sludge management and reagent costs. The operational costs of SBR technology are lower compared to the operating expenses of a conventional ASP (Dutta & Sarkar, 2015). The OPEX/CAPEX proportion of SBR technology was 21.8% (Ochs et al., 2023). However, the CAPEX and OPEX details of plants operation must be provided in sources for various IITs using the technology.

6.3 Moving Bed Biofilm Reactor

Moving bed biofilm reactors utilise biofilm carriers that reduce CAPEX and OPEX costs. The investment cost of MBBR technology ranges from moderate to high, considering India's standard land area requirements (Singh et al., 2018). Further, in a case study, Singh et al. (2018) analysed an institutional MBBR facility designed for 2-MLD and 0.4-MLD plants in Mumbai and Chennai, respectively. The authors highlighted that for the Chennai plant, the proportion of OPEX/CAPEX was 50.6%. The only difference between the two plants was the presence of an aeration tank prior to MBBR, and ozone as a disinfection system in the case of Mumbai plant. Many consultancy firms working on SBR and MBBR technologies emphasised that the percentage of hardware and civil costs, on average, was the same for both plants.

Alternatively, in a case study done by Haz Hazarika and Pandit (2018), a comparative economic estimation for a 0.1-MLD residential wastewater treatment plant operated with MBBR technology had a CAPEX and OPEX of INR 16 lakhs and INR 1,50,000 per month, respectively. The 0.005-MLD SBR residential plant had a CAPEX and OPEX of INR 16 lakhs and INR 10,000 per month. As both cases involved small residential areas with a population of 500–1,000, it could be generalised to compare higher educational institutions described here. SRM Chennai has three treatment plants based on MBBR technology with water demand of 1.6-, 2.5-, and 3.7-MLD.

Hence, considering similarities in scale, a rough CAPEX estimate of the three plants, assuming linearity, is presented in Table 3. It can be easily observed that CAPEX increases linearly with increasing capacity. Considering SBR technology, its rough CAPEX values are presented in Table 4. Although the estimates are based on assumptions, a higher CAPEX value for 3-MLD IIT Roorkee (n.d.) plant could be due to the hilly terrain that might escalate engineering costs.

6.4 Activated Sludge Process

A study conducted by Yang and researchers in 2020 discussed the suitability of textile industry wastewater via utilisation of ASP, MBR, and MBBR technologies. The results suggested

Table 4 Probable CAPEX values of SBR plants installed at IITs, considering linearity of scales as per Haz Hazarika and Pandit (2018)

S. no	Sewage treatment plant (SBR)	CAPEX (INR)
1	IIT Madras, Chennai (4-MLD)	12,80,00,000
2	IIT Roorkee, Uttarakhand (3-MLD)	9,60,00,000
3	IIT Mandi, Himachal Pradesh (0.450 MLD)	1,44,00,000
4	IIT Mandi, Himachal Pradesh (0.025 MLD)	80,00,000
5	IIT Mandi, Himachal Pradesh (0.125 MLD)	4,00,00,000

that the operating expenses (OPEX) of MBBR were comparable to those of MBR, and both were lower than the OPEX of ASP system. However, MBBR exhibited reduced investment costs and CAPEX, demonstrating a 68% decrease, compared to the CAPEX of MBR and ASP.

7 Need for Higher Education Institutions to Publish Information Regarding Sewage Treatment Plants

(a) **Transparency and accountability**: Every HEI is committed to manage solid and liquid wastes against the backdrop of impending SDGs. An online presentation of technological prowess of on-campus wastewater management shows institutional commitment to improving environmental practices and wastewater management by providing information about STPs.

(b) **Environmental education**: Practices, such as zero liquid discharge, support responsible wastewater management and circular economy. Additionally, an on-campus STP offers educational opportunities for both students and visitors. Students learn about environmental science, engineering, and sustainability through the practical examples of wastewater treatment processes. The research and development and the overall functioning give a clearer idea about the on-field success of such technologies. Displaying information about STPs leaves a lasting educational legacy. In addition to influencing their behaviours and decisions, it imparts the values of environmental responsibility to present and future generations.

(c) **Regulatory compliance and social responsibility**: In many regions, institutions are bound by regulations to operate wastewater treatment facilities that adhere to established environmental standards. Sharing details of these facilities can visibly demonstrate the institution's commitment to regulatory requirements. Moreover, most of these institutions are located at remote locations, especially in rural areas, which lack basic sewage treatment facilities. Hence, including those sewer lines in the institute's STP increases the overall treatment efficiency of the complete area.

(d) **Resilience education**: A yearly report of total energy and cost savings because of the operation of the plant helps to understand the long-term techno-economic viability of the plant, which can help in better financial decision-making. Furthermore, design robustness and operational continuity during non-stationary climatic extremes reinforce a massive win for the institute's sustainable infrastructure.

8 Conclusions

The innovative and sustainable approaches adopted for wastewater management can be discovered in the sewage treatment methods used by some of India's premier HEIs. By combining various sewage treatment methods from conventional to state-of-the-art technologies, these institutes are not only attending to their immediate wastewater management requirements but also actively promoting more significant environmental preservation and public health enhancement objectives. Currency of modern wastewater treatment technologies, such as MBBR and SBR, should be positively included in the wastewater management infrastructure considering the techno-economic upsides, compared to the conventional ASP. As India progresses on its path towards sustainable development, these HEIs assume a crucial role in establishing standards for conscientious sewage treatment, shaping policy dialogues, and championing the cause of efficient environmental governance, leading to a better decentralised sewage treatment framework.

References

Abhinav, A. (2015). Single-stage anaerobic digestion for sewage sludge treatment. B. Tech final year project report, National Institute of Technology Rourkela, Jharkhand.

Aerofloat. (2022). Sewage technology for university. Available at: https://www.aerofloat.com.au/sewage-technology-for-university-2/. Accessed on 7 Aug 2023

Amity University, Gurugram, Haryana. (n.d.). SDG 6: Clean water & sanitation (PDF file). Available at: https://www.amity.edu/gurugram/sdg/sdg17/sdg%2006%20clean%20water%20&%20sanitation.pdf

Bhargava, A. (2016). Activated sludge treatment process—Concept and system design. *International Journal of Environmental Development and Research, 4*(2), 890–896.

Bolong, N., Sabli, M. Q. N., Saad, I., & Ali, A. N. A. (2022). The efficiency of sewage treatment plant: A case study at the main campus of University Malaysia Sabah (UMS). *IOP Conference Series: Materials Science and Engineering, 1229*(1), Article 012013. https://doi.org/10.1088/1757-899X/1229/1/012013

Central Pollution Control Board (CPCB). (2021). National inventory of sewage treatment plants. Available at: https://cpcb.nic.in/openpdffile.php?id=UmVwb3J0RmlsZXMvMTIyOF8xNjE1MTk2MzIyX21lZGlhcGhvdG85NTY0LnBkZg

Centre for Science and Environment (CSE). (n.d.). Constructed wetland to treat wastewater at Indian Institute of Technology Powai, Mumbai. Available at: https://www.cseindia.org/constructed-wetland-to-treat-wastewater-at-indian-institute-of-technology-powai-mumbai-6209

Chatterjee, B. (2016). IIT-Bombay professor leads the way to save water. *Hindustan Times.* Available at: https://www.hindustantimes.com/mumbai/iit-bombay-professor-leads-way-to-save-water/story-iCsJBZU0bYbz3FJtI4PbiP.html

Cross, K., Tondera, K., Rizzo, A., Andrews, L., Pucher, B., Istenič, D., Karres, N., & Mcdonald, R. (2021). *Nature-based solutions for wastewater treatment*. IWA Publishing.

Das, S., & Bokshi, S. (2017). Sustainable wastewater treatment in developing countries: A case study of IIT Kharagpur campus. *Journal of the Institution of Engineers (India): Series A, 98*(1–2), 127–134. https://doi.org/10.1007/s40030-017-0197-z

Dutta, A., & Sarkar, S. (2015). Sequencing batch reactor for wastewater treatment: Recent advances. *Current Pollution Reports, 1*(3), 177–190. https://doi.org/10.1007/s40726-015-0016-y

Eriks, T. (2011). Linda Grinberga energy-efficient wastewater treatment technologies in constructed wetlands. In *Proceedings of the 3rd International Conference Civil Engineering'11 Proceedings V Environmental Engineering, Bangalore, India* (pp. 263–266).

Haz Hazarika, S., & Pandit, D. (2018). Feasibility Analysis of Decentralized Wastewater Treatment Technology: SBR and MBBR. *International Journal of Advances in Engineering Science and Technology, 6*(3), 23–26. http://iraj.in

IIT Mandi. (n.d.). Sewage treatment plant infrastructure. Available at: https://infra.iitmandi.ac.in/stp.php

IIT Roorkee. (n.d.). Sewage treatment plant. Indian Institute of Technology Roorkee. Available at: https://www.iitr.ac.in/estateworks/stp.html

Jones, E. R., van Vliet, M. T. H., Qadir, M., & Bierkens, M. F. P. (2021). Country-level and gridded estimates of wastewater production, collection, treatment, and reuse. *Earth System Science Data, 13*(2), 237–254. https://doi.org/10.5194/essd-13-237-2021

Krishnamurthy, K. (2019). IIT Madras—STP at IIT Madras 2019 Prof Ligy & Com. Video YouTube. Available at: https://www.youtube.com/watch?v=_nHu64z_Jlk

Kuo, H.-W.D., Zure, D., & Lin, C.-R. (2023). Occurrences of similar viral diversity in campus wastewater and reclaimed water of a university dormitory. *Chemosphere, 330*, Article 138713. https://doi.org/10.1016/j.chemosphere.2023.138713

Langergraber, G., Pressl, A., Kretschmer, F., & Weissenbacher, N. (2018). Small wastewater treatment plants in Austria—Technologies, management, and training of operators. *Ecological Engineering, 120*, 164–169. https://doi.org/10.1016/j.ecoleng.2018.05.030

LocalWiki. (n.d.). UC Davis wastewater treatment plant. Available at: https://localwiki.org/davis/UC_Davis_Wastewater_Treatment_Plant. Accessed on 7 Aug 2023

Mahatma Gandhi National Council for Rural Education (MGNCRE). (2019). *Jal Shakti Campus and Jal Shakti Gram.*

Masi, F., Bresciani, R., Martinuzzi, N., Cigarini, G., & Rizzo, A. (2017). Large-scale application of French reed beds: Municipal wastewater treatment for a 20,000 inhabitant's town in Moldova. *Water Science and Technology, 76*(1), 134–146. https://doi.org/10.2166/wst.2017.188

Masoud, A. M. N., Alfarra, A., & Sorlini, S. (2022). Constructed wetlands as a solution for sustainable sanitation: A comprehensive review on integrating climate change resilience and circular economy. *Water (Switzerland), 14*(20), 3232. https://doi.org/10.3390/w14203232

Negi, R., & Chandel, M. K. (2021). Analyzing water-energy-GHG nexus in a wastewater treatment plant of Mumbai Metropolitan Region, India. *Environmental Research, 196*, 110931. https://doi.org/10.1016/j.envres.2021.110931

Ochs, P., Martin, B., Germain-Cripps, E., Stephenson, T., van Loosdrecht, M., & Soares, A. (2023). Techno-economic analysis of side stream ammonia removal technologies: Biological options versus thermal stripping. *Environmental Science and Ecotechnology, 13*, Article 100220. https://doi.org/10.1016/j.ese.2022.100220

Parde, D., Patwa, A., Shukla, A., Vijay, R., Killedar, D. J., & Kumar, R. (2021). A review of constructed wetland on type, treatment, and technology of wastewater. *Environmental Technology & Innovation, 21*, Article 101261. https://doi.org/10.1016/j.eti.2020.101261

Parikh, A., Srivastava, L. P., Jain, S. K., Harish, P. M., Greene, M., Deth, V., Bhandarkar, S., & Kethineedi, M. (2019). *Water and wastewater management*. Indian Institute of Technology Gandhinagar.

Pennsylvania State University. (2023). *Wastewater treatment at Pennsylvania State University*. Available at: https://www.opp.psu.edu/unit/wastewater. Accessed on 7 Aug 2023

Singh, P., & Kansal, A. (2018). Energy and GHG accounting for wastewater infrastructure. *Resources, Conservation and Recycling, 128*, 499–507. https://doi.org/10.1016/j.resconrec.2016.07.014

Singh, A., Kamble, S. J., Sawant, M., Chakravarthy, Y., Kazmi, A., Aymerich, E., Starkl, M., Ghangrekar, M., & Philip, L. (2018). Technical, hygiene, economic, and life cycle assessment of full-scale moving bed biofilm reactors for wastewater treatment in India. *Environmental Science and Pollution Research, 25*(3), 2552–2569. https://doi.org/10.1007/s11356-017-0605-y

SRM Institute of Science and Technology (SRMIST). (2021). SRM sewage treatment plant report. *SRM Institute of Science and Technology*. Available at: https://webstor.srmist.edu.in/web_assets/downloads/2021/srm-sewage-treatment-plant-report.pdf

Statista. (2021). Number of higher education institutions in India 2021, by type. Available at: https://www.statista.com. Accessed on 6 Aug 2023

University of Stuttgart. (n.d.). Wastewater treatment plant. Available at: https://www.iswa.uni-stuttgart.de/en/wastewater-treatment-plant/plant/. Accessed on 7 Aug 2023

Vellore Institute of Technology (VIT) University. (n.d.). Liquid waste management. Available at: https://vit.ac.in/liquid-waste-management. Accessed on 6 Aug 2023

World Economic Forum (WEF). (2016). How the world's top universities are distributed. Available at: https://www.weforum.org/agenda/2016/10/top-universities-distributed-world-map/

Yang, X., López-Grimau, V., Vilaseca, M., & Crespi, M. (2020). Treatment of textile wastewater by CAS, MBR, and MBBR: A comparative study from technical, economic, and environmental perspectives. *Water, 12*(5), 1306. https://doi.org/10.3390/w12051306

Assessment of Drinking Water Treatment Plant Efficiency and Water Quality Through Indexing Approach

Sumona Koley, Meena Khwairakpam, and Ajay S. Kalamdhad

1 Introduction

Water is a necessity of life. It is one of the earth's most significant natural resources (Desye et al., 2021). Although it makes up for more than 71% of the Earth's total surface, it is always in short supply. The United Nations research indicates that by 2025, the current consumption patterns could result in water scarcity for almost one-third of the global population (Awotwi et al., 2019; Nivesh et al., 2023). In recent years, the water environment has been significantly impacted by industry and agriculture in different ways, making it sensitive to any local disturbances regardless of its location (Edokpayi et al., 2018; Mishra et al., 2024). Several factors, such as infringement, toxic waste from industrial and urban sources, silt, excessive pesticide nutrient addition, unrestricted use of fertilisers, human waste disposal, storm and agricultural runoff, and organic and inorganic waste materials, disturb the environment (Ramya et al., 2022).

The primary objective of drinking water treatment plants (DWTPs) is to ensure the provision of clean and safe water. They are made up of several physical and chemical barriers that protect against pathogens and potentially dangerous substances in raw water. The main objectives of DWTPs are to remove pollutants and pathogenic bacteria from freshwater while also making it safe to consume in terms of colour, odour and taste (Acarer, 2023). The intended usage of water drives the designing of the water treatment. Prior to its distribution to customers, water must undergo treatment to achieve the appropriate levels of physico-chemical properties while also being free from bacteria and pathogens. Each country has its own set of laws to check and control the purity of water. The World Health Organization (WHO, 2022) has established the most general regulations. Coagulation, sedimentation, filtration and disinfection are some of the conventional treatment units utilised during the treatment of water. Water treatment often integrates physical, biological and chemical processes. Advanced technologies are typically used when conventional methods work inefficiently (for instance, in the case of extremely contaminated water) (Grzegorzek et al., 2023).

The northeastern area of India is endowed with a vast interconnected system of rivers and lakes as well as an abundance of water resources. Nearly all Indian rivers are contaminated by the discharge of raw, untreated or just minimally processed wastewater and discharged industrial waste, and the northeast areas are no exception (Goswami & Kalamdhad, 2022). Even after the prevalence of pollution sources and the enormous illness burden linked to unsafe water sources, limited research has been done to evaluate the performance of current treatment plants and drinking water quality in low-income nations. Therefore, this study may have its implications for corrective action from water supply authorities, policymakers and decision-makers. The study aimed at determining the performance of water treatment plants and the quality of drinking water offered by these treatment plants operating in Guwahati, Assam, India.

S. Koley (✉) · A. S. Kalamdhad
Centre for the Environment, Indian Institute of Technology Guwahati, Guwahati, Assam, India
e-mail: sumonakoley95@gmail.com

M. Khwairakpam · A. S. Kalamdhad
School of Agro and Rural Technology, Indian Institute of Technology Guwahati, Guwahati, Assam, India

A. S. Kalamdhad
Department of Civil Engineering, Indian Institute of Technology Guwahati, Guwahati, Assam, India

© Capital Publishing Company, New Delhi, India 2025
I. Haq et al. (eds.), *Water and Wastewater*, https://doi.org/10.1007/978-3-031-94253-2_11

2 Materials and Methods

2.1 Study Area

For the study, five water treatment plants were chosen in and around the Guwahati city of Assam. Water samples were collected for analysis from the inlet and outlet of every treatment plant. The details of locations and other descriptions are shown in Fig. 1 and Table. 1. All water treatment facilities mostly use conventional water treatment procedures, which include aeration, coagulation followed by flocculation, rapid sand filtration and chlorination, imitating the vast majority of India's waterworks in the process.

1. Panbazar DWTP (TP1) (26.18768 N, 91.74314 E)
2. North Guwahati Water Supply (TP2) (26.19148 N, 91.72293 E)
3. IIT Guwahati DWTP (TP3) (8 MLD) (26.19202 N, 91.6946 E)
4. Jalukbari DWTP (TP4) (26.14809 N, 91.67128 E)
5. GMDA Water supply project (TP5) (26.14237 N, 91.65395 E)

Three treatment plants (TP1, TP2 and TP4) are currently operated by the public health engineering department (PHED) of Assam. TP3 is monitored by IIT Guwahati, and the Guwahati Metropolitan Development Authority (GMDA) monitors TP5, which is situated at Sadilapur, Kamrup rural district. Required permissions were taken from the concerned departments prior to starting sampling from treatment plants.

2.2 Sampling Protocol and Analytical Methods

The raw (inlet) and treated water samples (supply water) were collected in triplicate by following the grab sampling method from each DWTP in a dry and clean 1-L plastic container for laboratory analysis. Containers were washed thrice with water sample prior to sampling. The latitude and longitude of the sampling sites were located using Global Positioning System (GPS). Samples were labelled, sealed and brought to the laboratory in an icebox (4 °C) and stored in a refrigerator at a temperature of 4 °C. The sampling process was carefully carried out by adhering to the American Public Health Association's (APHA) recommended procedures. Water samples were examined in the laboratory for the following parameters: pH, electrical conductivity (EC), turbidity, alkalinity, hardness, chlorides, fluorides, nitrates, phosphates and sulphates.

Fig. 1 Locations of drinking water treatment plants

Table 1 Operational parameter of the DWTPs studied

Plant operational parameters	TP1	TP2	TP3	TP4	TP5
Plant capacity (MLD)	12.15	4.5	8.4	4	53.5
Plant location	Panbazar	Amingaon	IITGuwahati	Jalukbari	Sadilapur
Influent water	Brahmaputra	Brahmaputra	Brahmaputra	Brahmaputra	Brahmaputra
Treatment	Conventional	Conventional	Conventional	Conventional	Conventional
Pre settling tank	×	×	✓	×	✓
Coagulant used	Alum	Alum	Alum	Alum	Alum
Flash mixer	×	×	✓	×	✓
Disinfectant used	Ca(OCl)Cl	Ca(OCl)Cl	NaCl	Ca(OCl)Cl	NaOCl

2.3 Water Quality Index

An efficient and reliable method for assessing the overall quality of water is water quality index (WQI). It is a highly used method of averaging data on several aspects of consolidated water quality into a singular outcome. The WQI approach offers decision-makers and other stakeholders valuable insights into the comprehensive water quality assessment (Rakotondrabe et al., 2018). Each parameter's load was determined by considering its importance for drinking and its perceived impact on human health. The WQI was computed using a total of 10 parameters, each of which was given a variable weightage (W_A) based on relative impact on the overall quality of water, which ranged from 1 to 5 (Table 2). Categories with the least impact on water quality were given weights of 5 and 1, respectively. To determine relative weights (W_R), the following equation was used:

$$W_R = \frac{W_A}{\sum W_A} \tag{1}$$

Table 2 Relative weights of water quality parameters

Parameters	Weight (W_A)	Relative weight (W_R)
pH	2	0.05
EC	2	0.05
Alkalinity	2	0.05
Hardness	3	0.075
Fluoride	4	0.1
Chloride	4	0.1
Turbidity	5	0.125
Nitrate	5	0.125
Phosphate	5	0.125
Sulphate	5	0.125

where W_A is the weight assigned to each parameter, and W_R is the relative weight of each individual parameter. Along with individual parameter weight, W_R of each parameter is displayed in Table 2.

The quality rating scale, Q_i was determined by dividing each parameter's concentration by the Bureau of Indian *Standards* (BIS) permissible limit (BIS 10500: 2012):

$$Q_i = \frac{C_i}{S_i} \times 100 \tag{2}$$

where C_i is the concentration of each parameter in each sample and S_i is the corresponding standard value of that parameter provided by BIS 10500:2012.

The quality scaling rate of pH was calculated using different methods. The observed values of the samples were subtracted from the ideal value, (V_i) of pH 7 (Alobaidy et al., 2010).

For pH,

$$Q_i = \frac{C_i - V_i}{S_i - V_i} \times 100 \tag{3}$$

The sub-index, SI_i for each parameter was calculated using the following equation:

$$SI_i = W_A \times Q_i \tag{4}$$

where W_R is the relative weight calculated using Eq. (1) and shown in Table 2.

Q_i is the quality rating scale calculated using Eq. (2)

Water Quality Index, WQI $= \sum SI_i$.

2.4 Data Quality and Assurance

Aseptic techniques, control medium, blank measurements, correct equipment sterilisation, triplicate analysis and aseptic processes were all used to maintain the data's quality. Additionally, standard methods of analysis and sampling were used.

2.5 Analysis and Management of Data

Prior to entering raw data into an Excel 2021 spreadsheet, each set of data was coded. Standard deviations and mean values were computed. The overall removal efficiency of the treatment plant was evaluated using the following equation:

$$\text{Removal efficiency }(\%) = \frac{C_i - f_i}{C_i} \times 100$$

where C_i is the initial pollutant concentration and C_f is the final pollutant concentration after treatment.

3 Results and Discussion

3.1 Physico-chemical Properties of Raw and Treated Water

It has been proven that extended exposure to physico-chemical pollutants in drinking water can have a negative impact on human health. Both organic and inorganic compounds are included in this category. Certain pollutants are detrimental to human health and diminish the overall visual appeal of water. Physico-chemical parameter analysis of raw water and treated water of five treatment plants are reported in Table 3.

For all treatment plants, water is supplied from the river Brahmaputra. The pH of water affects both chemical and biological processes taking place in aquatic environments. Because photosynthesis consumes carbon dioxide (CO_2) and causes pH to rise, the growth of aquatic plants may also contribute to an increase in the river's pH (Azzam et al., 2022). Because of the high flow of water in the river, various activities of aquatic plants are not discovered. Results of the research showed that all pH values of water samples collected from either inlets or outlets were within the acceptable range of 6.5–8.5; final pH values were in the range of 6.5–7.8.

Ion dissociations, temperature and electrical field mobility speed determine EC. The environment, water movement and water supply determine ion concentrations. Surface water contains soluble ions, mostly from rock material solutions. The EC of inlet water samples ranged between 110 μS/cm and 139 μS/cm. TP2 received water with the maximum EC, and TP3 received water with the lowest EC. The EC of treated water samples ranged between 123 μS/cm and 149 μS/cm.

Alkalinity is a metric used to determine how well water can neutralise acids. Water's acidity is reduced, and the H^+ is successfully removed by alkalinity brought on by the presence of HCO_3^-, CO_3^{2-} and OH^-. It functions as a buffer and aids in preserving the pH equilibrium of water. Rocks containing CO_3^{2-}, HCO_3^- and OH^- are notable examples of the natural sources of alkalinity. The river water alkalinity ranged well below the permissible value. A slight decrease in water alkalinity was seen after treatment.

The same trend followed in the case of water hardness. Total hardness (TH) is the measure of the presence of multivalent cations in water, mostly Ca^{2+} and Mg^{2+}. Seepage, runoff and dissolved metallic ions from sedimentary rocks are the primary natural sources of total hardness in water. Although hard water has no harmful impacts on human health, it has a significant detrimental effect on boilers, cooling towers, and other industrial equipments that hold or process water.

One of the crucial and determining factors for the usability of drinking water is turbidity. The raw water has high level of turbidity in the range of 70–80 nephelometric turbidity units (NTU). Turbidity of the treated water was within limits, except for TP1 and TP2, where the turbidity levels in treated water samples were >2 NTU.

Anions, such as fluorides, chlorides, nitrates, phosphates and sulphates, were present in raw water. Except for phosphates, other anions with very less concentration were discovered in raw water. The inlet water sample of TP5 had no phosphates. In other samples, the phosphate concentration was discovered above the permissible limit. After treatment, level of anions further decreased in water, except chlorides and sulphates, which increased manifold after the treatment.

3.2 Removal Efficiency of Different Pollutants Through Conventional Drinking Water Treatment Plants

The results of removal efficiencies of physical and chemical parameters are shown in Table 4. pH is a significant aspect in determining whether water is suitable for a given purpose. In case of all treatment plants, the pH of water decreased slightly after conventional treatment. Compared to the pH of inlet water samples, decrease in the pH of treated water ranged from 6 to 11%. The decrease in pH could be due to the removal of alkalinity through treatment caused by the bicarbonates present in water. The EC value increased after the treatment of water in every treatment plant, except TP2, which received water with maximum EC value. In other plants, the EC value in water increased by 5–11%. The maximum increase of 11.31% was observed in TP3. Addition of different chemicals throughout the treatment processes could increase the EC of treated water.

The alkalinity of raw water was slightly removed through the conventional water treatment processes, except TP1, where the alkalinity of water after treatment increased by

Table 3 Physico-chemical parameter analysis of raw water and treated water of five DWTPs

Parameters	TP1		TP2		TP3		TP4		TP5	
	C_i	C_f	C_i	C_f	C_i	C_f	C_i	C_f	C_i	C_f
pH	7.5 ± 0.24	6.83 ± 0.04	7.8 ± 0.20	6.93 ± 0.14	7.24 ± 0.24	6.57 ± 0.17	7.18 ± 0.02	6.65 ± 0.22	7.02 ± 0.03	6.57 ± 0.14
EC (µS/cm)	129 ± 7.77	136 ± 2.83	147 ± 4.95	142 ± 6.36	110 ± 0.71	123 ± 1.41	121 ± 4.95	133 ± 2.12	139 ± 2.83	149 ± 6.36
Alkalinity (mg/L as $CaCO_3$)	56 ± 2.83	58 ± 2.12	57 ± 3.54	51 ± 3.54	57 ± 1.70	47 ± 2.12	59 ± 2.82	50 ± 2.64	55 ± 3.54	49 ± 3.33
Hardness (mg/L as $CaCO_3$)	62 ± 3.05	48 ± 2.41	60 ± 3.32	55 ± 1.87	58 ± 2.88	54 ± 1.65	59 ± 4.56	49 ± 1.76	59 ± 2.33	54 ± 0.59
Fluorides (mg/L)	0.161 ± 0.04	0.16 ± 0.02	0.156 ± 0.03	0.149 ± 0.06	0.168 ± 0.06	0.139 ± 0.02	0.161 ± 0.02	0.136 ± 0.04	0.157 ± 0.02	0.138 ± 0.07
Chlorides (mg/L)	1.974 ± 0.82	6.578 ± 2.01	1.841 ± 0.76	3.94 ± 1.05	5.137 ± 1.88	4.606 ± 2.07	2.291 ± 0.84	3.287 ± 1.30	2.989 ± 0.98	5.739 ± 1.88
Turbidity (NTU)	80 ± 4.22	3 ± 0.23	64.5 ± 2.50	4.5 ± 0.44	70 ± 5.98	1.5 ± 0.08	75.5 ± 6.03	1.5 ± 0.15	77 ± 6.12	1 ± 0.05
Nitrates (mg/L)	1.818 ± 0.42	0.704 ± 0.03	1.312 ± 0.17	0.629 ± 0.06	5.547 ± 1.01	2.706 ± 0.98	1.769 ± 0.92	2.125 ± 1.04	1.394 ± 0.28	0.842 ± 0.02
Phosphates (mg/L)	0.158 ± 0.05	0.128 ± 0.07	0.221 ± 0.03	0.177 ± 0.03	0.309 ± 0.01	0.104 ± 0.05	0.108 ± 0.06	0.092 ± 0.07	ND	ND
Sulphates (mg/L)	24.746 ± 2.98	32.041 ± 4.75	24.357 ± 3.01	32.219 ± 2.54	24.911 ± 1.78	36.335 ± 4.47	26.159 ± 3.32	47.059 ± 5.51	25.027 ± 2.78	33.18 ± 3.54

Note NTU: nephelometric turbidity units

Table 4 DWTPs' removal efficiency (%) for selected parameters

	TP1	TP2	TP3	TP4	TP5
pH	9	11.15	9.26	7.39	6.41
EC	−5.42	3.40	−11.31	−9.92	−7.19
Alkalinity	−3.57	10.53	17.54	15.25	10.91
Hardness	22.58	8.33	6.90	16.95	8.47
Fluorides	0.62	4.49	17.26	15.53	12.10
Chlorides	−233.23	−114.01	10.34	−43.47	−92.00
Turbidity	96.25	93.02	97.86	98.01	98.70
Nitrates	61.28	52.06	51.22	−20.12	39.60
Phosphates	18.99	19.90	66.34	14.81	ND
Sulphates	−29.48	−32.28	−45.86	−79.90	−32.58

3.57%. Other four treatment plants removed alkalinity by 10–17%. TP3 removed the maximum alkalinity of 17.54%. Regarding the removal of water hardness, TP1 was the most efficient facility in removing water hardness (22.58%), followed by TP4 (16.95%).

Colloidal particles play an important role in water chemistry. The presence of colloidal solids in water creates turbidity. It was found in the present study that the turbidity removal efficiency of every treatment plant was >90%, with the highest removal efficiency observed in TP3, TP4 and TP5, which removed around 98% of the turbidity through conventional treatment.

For the removal of anions from raw water, all treatment plants worked efficiently, except for sulphate and chloride ions, which increased hugely after the treatment. This could be because of the usage of different chemicals throughout treatment processes. Similar results were determined by Mohammed et al. (2022).

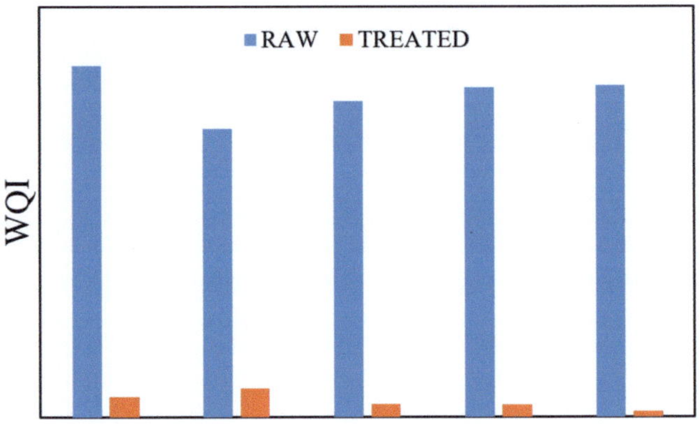

Fig. 2 Weighted arithmetic WQI for raw and treated water samples

Table 5 Classification of water quality (Yadav et al., 2010)

WQI	Water quality rating
0–25	Excellent
26–50	Good
51–75	Poor
76–100	Very poor
>100	Unsuitable for drinking

3.3 Water Quality Index Prior to and After Water Treatment

To provide an overview of the overall condition of drinking water's quality, the WQI was calculated for both raw water and treated water. The weighted arithmetic WQI of water is shown in Fig. 2. As depicted in Table 5, the WQI of river water from all treatment plants demonstrated that water was unsuitable for drinking purposes, and the results mandated the obvious necessity for treatment before supplying water to consumers.

On the other hand, although the WQI for treated water samples was within the range, huge variation was observed in water quality of five treatment plants. The water from TP5 had the best water quality as its WQI was in the range of excellent water quality category. TP3 and TP4 had good quality of treated water, but it was not excellent, probably because turbidity level was more than the permissible limit, and the presence of phosphates in water was also above the permissible limit. The water quality of TP1 was poor because of its inefficiency in removing the turbidity to the desired limit and also presence of high phosphate concentration. For the same reason, TP2 also provided water of very poor quality, compared to the water of the other four TPs.

4 Conclusions

The results showed that several water quality parameters were above the BIS standard values proposed for drinking water quality. The removal effectiveness of treatment facilities was uncertain, and the WQI was rated outstanding for TP5, fair for TP3 and TP4, poor for TP1, and very poor for TP2. To adequately treat drinking water and increase the effectiveness of treatment plants, it is necessary to perform an efficient preliminary treatment, such as screening to lower the incoming organic load, proper chlorination of the drinking water system, and routine monitoring and maintenance of the treatment plant system. Additionally, steps must be taken to limit the amount of phosphates in both untreated and treated water. DWTPs must be present for the treatment of water prior to its distribution to consumers. Additionally, treatment methods must be adjusted periodically to account for the sharp changes in the pollution levels of raw surface water.

References

Acarer, S. (2023). Abundance and characteristics of microplastics in drinking water treatment plants, distribution systems, water from refill kiosks, tap waters and bottled waters. *Science of the Total Environment, 884*, Article 163866.

Alobaidy, A. H. M. J., Haider, S. A., & Bahram, K. M. (2010). Application of water quality index for assessment of Dokan lake ecosystem, Kurdistan region, Iraq. *Journal of Water Resource and Protection, 2*(9), 29093. https://doi.org/10.4236/jwarp.2010.29093

Awotwi, A., Anornu, G. K., Quaye-Ballard, J. A., Annor, T., Forkuo, E. K., Harris, E., Agyekum, J., & Terlabie, J. L. (2019). Water balance responses to land-use/land-cover changes in the Pra River Basin of Ghana, 1986–2025. *CATENA, 182*, Article 104129. https://doi.org/10.1016/j.catena.2019.104129

Azzam, M. I., Korayem, A. S., Othman, S. A., & Mohammed, F. A. (2022). Assessment of some drinking water plants efficiency at El-Menofeya Governorate, Egypt. *Environmental Nanotechnology, Monitoring & Management, 18*, 100705. https://doi.org/10.1016/j.enmm.2022.100705

Desye, B., Belete, B., Asfaw Gebrezgi, Z., & Terefe Reda, T. (2021). Efficiency of treatment plant and drinking water quality assessment from source to household, Gondar city, Northwest Ethiopia. *Journal of Environmental and Public Health,* 1–8. https://doi.org/10.1155/2021/9974064

Edokpayi, J. N., Odiyo, J. O., Popoola, E. O., & Msagati, T. A. (2018). Evaluation of microbiological and physicochemical parameters of alternative source of drinking water: A case study of Nzhelele river, South Africa. *The Open Microbiology Journal, 12*, 18. https://doi.org/10.2174/1874285801812010018

Goswami, A. P., & Kalamdhad, A. S. (2022). Trend of Water quality of a river flowing through densely populated area of Assam, India using Water Quality Index (WQI). Haq, Kalamdhad, Dash (Eds) In *Environmental degradation: Monitoring, assessment and treatment technologies* (pp. 215–224). Springer International Publishing.

Grzegorzek, M., Wartalska, K., & Kaźmierczak, B. (2023). Review of water treatment methods with a focus on energy consumption. *International Communications in Heat and Mass Transfer, 143*, Article 106674. https://doi.org/10.1016/j.icheatmasstransfer.2023.106674

Mishra, A. P., Singh, S., Jani, M., Singh, K. A., Pande, C. B., & Varade, A. M. (2024). Assessment of water quality index using analytic hierarchy process (AHP) and GIS: A case study of a struggling Asan River. *International Journal of Environmental Analytical Chemistry, 104*(5), 1159–1171.

Mohammed, I., Al-Khalaf, S. K. H., Alwan, H. H., & Naje, A. S. (2022). Environmental assessment of Karbala water treatment plant using water quality index (WQI). *Materials Today: Proceedings, 60*, 1554–1560. https://doi.org/10.1016/j.matpr.2021.12.065

Nivesh, S., Patil, J. P., Goyal, V. C., Saran, B., Singh, A. K., Raizada, A., Malik, A., & Kuriqi, A. (2023). Assessment of future water demand and supply using WEAP model in Dhasan River basin, Madhya Pradesh, India. *Environmental Science and Pollution Research, 30*(10), 27289–27302. https://doi.org/10.1007/s11356-022-24050-0

Rakotondrabe, F., Ngoupayou, J. R. N., Mfonka, Z., Rasolomanana, E. H., Abolo, A. J. N., & Ako, A. A. (2018). Water quality assessment in the Bétaré-Oya gold mining area (East-Cameroon): Multivariate statistical analysis approach. *Science of the Total Environment, 610*, 831–844. https://doi.org/10.1016/j.scitotenv.2017.08.080

Ramya, M., Elumalai, S., & Umamaheswari, A. (2022). Study of physicochemical and statistical analysis of water quality parameters in Veeranam Lake, Cuddalore district, Tamil Nadu, India. *Materials Today: Proceedings, 49*, 2934–2942. https://doi.org/10.1016/j.matpr.2021.11.354

World Health Organization (WHO). (2022). *Guidelines for drinking-water quality: Incorporating the first and second addenda.* World Health Organization.

Yadav, A. K., Khan, P., & Sharma, S. K. (2010). Water quality index assessment of groundwater in Todaraisingh Tehsil of Rajasthan state, India—A greener approach. *E-Journal of Chemistry, 7*(S1), S428–S432. https://doi.org/10.1155/2010/419432

Index

© Capital Publishing Company, New Delhi, India 2025
I. Haq et al. (eds.), *Water and Wastewater*, https://doi.org/10.1007/978-3-031-94253-2